The Brain Book

This book is known as the **BRAIN** Book

As a student of birth, be it as a birth assistant or midwife, there is so much to know in the beginning, and it most likely will not all reside in your brain at first. What you cannot keep in your head, you will find in this book. It is a resource to help you along the way as you learn to walk alongside women and midwives, serving each, learning about pregnancy birth and the postpartum period as you become a midwife yourself. This book is meant to be a workbook as well as a quick reference for beginning birth attendants.

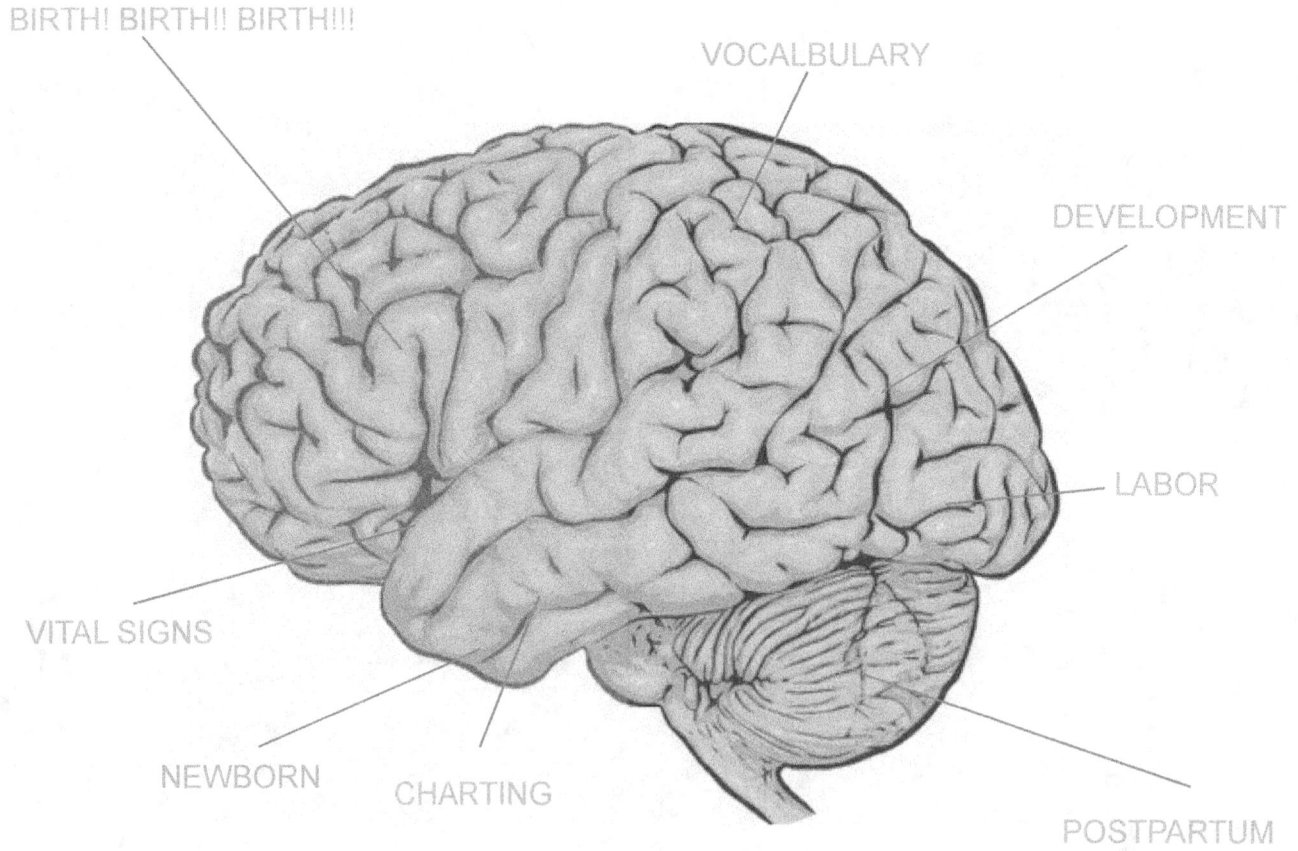

© 2016 GoMidwife. All rights reserved.
Pontus University Press

Table of Contents

Introduction to Midwifery Terminology...7
What is a Birth Assistant..37
The Scope of Birth Assistant..41
What is a Midwife...45
What is the Scope of a Midwife..45
Routes to Midwifery...46
The Spirits of Midwifery..48
Communication...51
Professionalism...55
Anatomy...57
Understanding the Menstrual Cycle...63
Fetal Development...66
    Week 1-2..66
    Week 3..67
    Week 4..68
    Week 5..69
    Week 6..70
    Weeks 7-9...70
    Weeks 10-12...71
    Weeks 13-16...72
    Weeks 17-20...73
    Weeks 21-28...74
    Weeks 29-34...74
    Weeks 35-38...75
    Weeks 39-42...76
The Function of Hormones..77
Nutrition...83
Normal Adult Vital Signs..89
    Blood Pressure..89
    Respirations..90
    Pulse...90
    Temperature...90
Routine Prenatal..92
Understanding Labs...103
Urinalysis...105
Blood Sugar Values...108
Fear in Pregnancy..109
Fetal Positions..111
Cephalic, Breech and Transverse..111
Fetal Positions per Vaginal Exam...112
Leopold's Maneuvers..114
Finding Fetal Heart Tones..115

When the BA Arrives... 116
Thinking on Your Feet...121
Signs of Labor...123
Normal Labor Patterns..124
Stages of Labor..125
Cervical Dilation..127
Station...128
Effacement...128
Charting...130
Times...132
APGAR...133
Vital Signs and Assessment...134
Newborn Exam...135
Reflexes...137
Ballard Gestational Exam..139
Newborn Exam SOAP...140
Pounds to Grams Conversion..142
Signs and Symptoms of Respiratory Distress..................................142
Inspecting the Placenta..143
Birth Note..146
Normal Newborn..147
 Temperature...147
 Respiratory Rate..147
 Heart Rate...147
 Weight...147
 Head Circumference..148
 Chest Circumference...148
 Length...148
 Blood Sugar...148
 Fontanelle..148
Signs and Symptoms of Infection..150
Normal Newborn Behavior..150
Common Newborn Skin Conditions...151
Normal Newborn Pee and Poop Chart...153
Lochia Chart...154
Postpartum Visit...156
Breastfeeding Assessment...158
Packing Your Birth Bag...160
End Notes..167

©2016 GoMidwife

Some day someone's life may depend on whether you know this, or not.
Debbie Perry, CPM

Introduction to Midwifery Terminology

The following abbreviations and terminology will help you as you enter into the field of midwifery. Midwifery has its own language and it is very important to understand new terms and to be able to break words down. Be sure to look up each abbreviation, root, suffix and word and write a thorough definition in the space provided below. Please note this is an overview of terms, not an exhaustive list. Space is provided for you to add terms during the course of your studies. To find the definition of terms and abbreviations simply type: midwifery terms and abbreviations into your Google search bar.

-cyesis	
-cyst	
-cyte	
-desis	
-dilation	
-ectomy	
-emia	
-ismus	
-itis	
-ology	
-ology	
-osis	
-oxia	
-pathy	
-pathy	
-penia	
-plasty	

-tomy	
A	
A, An	
AAT	
Abdomen	
Abdomin/o	
ABO	
Abortifacient	
Abruptio Placenta	
Accelerations	
Accretia	
Acidosis	
Active Phase of Labor	
After pains	
Alpha-Fetoprotein (AFP)	
Amenorrhea	
Amni/o	
Amnion	
Amniotic Fluid	
Anal Sphincter Muscles	
Anatomy	
Anemia	
Anencephaly	

Anomaly	
Antepartum	
Anterior	
Anterior Fontanelle	
Anterior Position	
Anti	
Antiseptic	
Anus	
Apgar Score	
Apnea	
Areola	
ARM	
Arteries	
Aspiration	
Asymptomatic	
Asynclitism	
Attitude	
Augment	
Axillary	
Axis of Birth Canal	

B	
Bacteria	
Bacterial Vaginosis	
Bacteriuria	
Ballotable	
Beta-strep	
Bi-	
Bicornuate	
Bilirubin	
Bimanual Compression	
Birth Canal	
Bladder	
Blanching	
Blasto-	
Blastocyst	
Blood Pressure	
Blood Sugar	
Bloody Show	
BOW/BBOW	
BP	
BPM	
Brachial	
Brady-	
Braxton Hicks	

Breech Position	
Breech Presentation	
Brown Fat	
C	
Candida	
Capillaries	
Caput Succedaneum	
Carcinogen	
Cardinal Movements	
Cardio-	
Carminative	
Catecholamine	
Catheterization	
CCK	
CCT	
Cell	
Cephalo	
Cephalic	
Cephalic	

Presentation	
Cephalohematoma	
Cephalopelvic Disproportion: CPD	
Cervic/o	
Cervical Block	
Cervical Cap	
Cervical Incompetence	
Cervical Mucus Method	
Cervical Os: internal and external	
Cervix	
Chlamydia	
Chromosomal Abnormalities	
Chronic	
Circumoral Cyanosis	
Cleft Lip (and Palate)	
Clitoris	
Clonus	
Coitus	
Coitus Interruptus	
Colic	

Colostrum	
Complication	
Congenital	
Contraception	
Contraction	
Contraindication	
Convulsion	
Cord Prolapse	
Crowning	
Ctx	
Cx	
Cyanosis	
Cyst (List types, causes and character of each.)	
Cystitis	
Cystocele	
Cyto-	
Cytomegalovirus	
D	
Dacryo-	

Decelerations	
Dehydration	
Denominator	
Descent	
Diabetes Mellitus	
Diastolic Pressure	
Dilation	
Discharge (List types, signs, symptoms and causes)	
Diuretic	
Doppler	
Doptone	
Doula	
Down's Syndrome (Trisomy 21)	
Dysfunction	
Dysmenorrhea	
Dyspnea	
Dystocia	
E	

Early Deceleration	
EBL	
Eclampsia	
Ectopic Pregnancy	
EDB	
EDD	
Edema	
Effacement	
EFW	
Embryo	
Emmenagogue	
Endo-	
Enema	
Engagement	
Engorgement	
Epi-	
Epi-	
Episiotomy	
Erythema Infectiosum	
Estrogen	
Exo-	
Expulsion	
Extension	
External Rotation	
External Version	

F	
Failure to Thrive	
Fallopian Tubes	
False Labor	
Feces	
Femoral	
Fertilization	
Fetal	
Fetal Axis Pressure	
Fetal Distress	
Fetal Lie	
Fetal Position	
Fetascaope	
Fetile Mucus	
Fetus	
FGM	
FHR	
FHT	
Fibroid	

First Stage of Labor	
Flexion	
Folic Acid	
Fontanelle	
Forewaters	
Friable	
Fundal Dominance	
Fundal Height	
Fundus	
Funic	
G	
Galactogogue	
Gastro-	
Gastrointestinal Tract	
GBS	
GC	
Genitals	

Genitourinary Tract	
Gestation	
Gestational Age	
Glucose	
Glyc/o-	
Gonorrhea	
Grand Multipara	
Gravida	
GTPAL	
GTT	
H	
Hem-	
Hematocrit	
Hemoglobin	
Hemorrhage	
Hemorrhoids (piles)	
Hernia	

Herpes	
Hindwaters	
Histio-	
Homan's Sign	
Homeopathy	
Hormones (List pregnancy, birth and postpartum hormones; purpose and function.	
Hx	
Hymen	
Hyper-	
Hyperpyrexia	
Hypertension	
Hyperventilation	
Hypo-	
Hypoglycemia	
Hypotensive	
Hypothermia	
Hypotonia	
Hypovolemic shock	
Hypoxia	
I	

Immunization	
Implantation	
Incompetent Cervix	
Incontinence	
Induction	
Inertia of Uterus	
Infant Mortality	
Infertility	
Inter-	
Intercostal	
Internal Rotation	
Internal Version	
Intra-	
Intrauterine	
Intravenous	
Involution	
Ischial Spines	
Ischial Tuberosity	
IUP	

J	
Jaundice	
K	
Ketonacidosis	
Ketone	
Ketonuria	
L	
Labia	
Labia Majora	
Labia Minora	
Lactation	
Lactobacillus	
Lanugo	
Late Deceleration	

Latent Phase of Labor	
Latero-	
Leukorrhea	
Lie	
Lightening	
Liquor	
LNPM	
LOA	
Lochia	
LOP	
LOT	
Lower Uterine Segment	
M	
Macronutrient	
Mal-	
Malaria	
Malnutrition	
Malpresentation	
Mammary Glands	

Mastitis	
Maternal	
Maternal Morbidity	
Maternal Mortality	
Mec	
Mechanism of Labor	
Meconium	
Menarche	
Menopause	
Menses	
Menstrual Cycle	
Menstrual Sponge	
Milia	
Miscarriage (List the types; signs and symptoms)	
Molding	
Mongolian Spot	
Mucous	
Mucous Membrane	
Multigravida	
Multipara	
Multiple Pregnancy	
Murmur	
My/o-	

N	
NAD	
Neonate	
Neuro-	
Neurological	
Nitrazine Test	
NML	
Nonspecific Vaginitis	
NSVD	
Nullipara	
O	
O/E	
OA	

Oblique Lie	
Obstetr/o	
Occiput	
Oligohydramnios	
Oliguria	
OP	
Osteo-	
Ovaries	
Ovulation	
Ovum	
Oxytocin	
P	
Palpate	
Pancreas	
Pap Smear	
Para	
Para-	
Parasite	
Parity	

Parous	
Parturient	
Patent	
Pathology	
Pelvic Exam	
Pelvic Floor	
Pelvic Inflammatory Disease (PID)	
Pelvimetry	
Pelvis	
Peri-	
Perinatal	
Perineal Muscles	
Perineum	
Physiological	
Physiological Retraction Ring	
Pic	
Pica	
Pitocin	
Placenta	
Placenta Accreta	
Placenta Previa	
Placental Insufficiency	
Polarity	

Poly-	
Polyuria	
Position	
Post Dates	
Post-	
Posterior	
Posterior Presentation	
Postpartum	
PP	
Pre-	
Precipitous Labor	
Predisposition	
Preeclampsia: Mild and Severe	
Pregnancy Induced Hypertension (PIH)	
Premature	
Premature Labor	
Prenatal	
Presentation	
Prima-	
Primagravida	
Primapara	
Progesterone	
Prolapse	

PROM	
Protein	
Proteinuria	
PTL	
Pubic Bone	
Pubic Mound	
Pulse	

Q

Quadrant	
Quickening	

R

Radial Pulse	
Rectocele	
Rectum	

Red Blood Cells	
Reflex	
Relaxin	
Renal	
Reproductive System	
Respiration	
Respiratory Distress Syndrome (RDS)	
Restitution	
Resuscitation	
Retained Placenta	
Retraction	
Retro-	
Retroverted	
Reynaud's Syndrome	
Rhesus Factor	
Rhogam	
Ripening	
Risk Factor	
ROA	
ROP	
ROT	
Rotation	
Round Ligament	

Rythm Method	
S	
SAB	
Sacrum	
Second Stage of Labor	
Secondary Powers	
Sepsis	
Serology	
Sexually Transmitted Disease	
Shock	
Shoulder Dystocia	
Sickle Cell Anemia	
Silver Nitrate	
Sitz Bath	
Souffle	
Specific Gravity	
Speculum	

Sperm	
Spina Bifida	
SROM	
Station	
Stripping Membranes	
STS	
Sub-	
Subinvolution	
Sucking Reflex	
Supine	
Suppository	
Suprapubic Pressure	
Surfactant	
Suture	
Syn-	
Syphilis	
Systemic	
Systolic Pressure	
T	

Tachy-	
Tachycardia	
Tachypneoa	
Tactile	
Teratogen	
Term	
Testicles	
Testosterone	
Third Stage of Labor	
Thrush	
Thyroid	
TOL	
Tone	
Torticalous	
Toxemia	
Toxo-	
Toxoplasmosis	
Transition	
Transverse Lie	
Trichomonas	
Trimester	
Trisomy 21	
True Labor	
Tubal Pregnancy	

U	
Ultrasound	
Umbilical Cord	
Umbilical Hernia	
Umbilicus	
Unstable Lie	
Upper Uterine Segment	
Urethra	
Urinalysis	
Urinary Incontinence	
Urinary Tract Infection (UTI)	
Urine Dip	
Uterine Ineria	
Uterus	

V	
Vacuum Extraction	
Vagal	
Vagina	
Vaginitis	
Variability	
Varicella	
Varicose Veins	
Vasa Previa	
Vaso-	
Vasodilation	
VBAC	
VE	
Velamentous	
Vena Cava	
Venereal Disease	
Vernix Caseosa	
Version	
Vertex	
Viable	
Virus	
Vital Signs	
Vital Statistics	
Vulva	

Vx	
W	
Webbed	
Well Baby Exam	
Wharton's Jelly	
White Blood Cells	
Witch's Milk	
WNL	
Womb	
X	
Xiphoid Process	
Y	
Yeast Infection	
Yolk Sac	

Z	
Zygote	

Notes _____
_____
_____
_____
_____
_____
_____
_____
_____
_____

Attending births is like growing roses. You have to marvel at the ones that just open up and bloom at the first kiss of the sun but you wouldn't dream of pulling open the petals of the tightly closed buds and forcing them to blossom to your time line.

Gloria Lemay

What is a Birth Assistant

A birth assistant serves not only a client in labor but the midwife who is in charge of her care. The role of a birth assistant may include doula support, prenatal and postpartum education, as well as assistance during the actual labor and delivery. Your role as a birth assistant will vary depending on the birth team you are working with as well as the circumstances surrounding the birth itself.

Important Principles of Birth Assisting

Serve according to your highest training and greatest ability. Do not step outside of the scope of your training unless directed and supervised by a trained professional.

Be flexible. Birth is not a scheduled or controlled event and it's important that you enter your role as assistant with the mindset that you are there to serve until you are no longer needed.

Be sacrificial. Considering the needs of others above yourself is not only a principal of an excellent birth assistant, but one we are called to as believers. You will likely encounter moments where you feel your personal strength is spent and you are ready to throw in the towel but it is at this moment when you are likely to be needed the most. Labor progresses over time and then reaches a pinnacle and your willingness to continue on will reap many lasting rewards. The Lord is gracious to us as we give of ourselves and He renews our strength when we ask.

Anticipate needs. As you begin to attend birth you will likely be able to anticipate the needs of those you work alongside. Begin to make mental (or physical!) notes and try to meet those needs before the midwife needs to ask. Don't worry, it takes time but as you learn the flow of birth, you will quickly begin to see ways that you can be helpful and take undue stress off of the rest of the birth team and most importantly, the laboring mom.

Home Birth Assistant in the West

As the assistant to a midwife at a home birth, you will be responsible for all of the details surrounding the birth so the midwife is able to give all of her attention to the laboring mother. Remember to be flexible. These are not a list of hard and fast tasks to check off. Watch for ways to serve & be willing to step in to fill needs as you're able. There are many facets to this role but you will most likely be asked to assist with the following:

- Set up/clean up- If you're attending a water birth, set up responsibilities may include the set up and tear down of a birth pool along with it's temperature maintenance (birth pools should always stay between 96-98°F or 34-37°C)
- Gathering pertinent information about the client and her status
- Communicating information to the midwife
- Keeping records/charting
- Birth supply set up- Ask the midwife you're working with how she likes supplies to be arranged. Knowing where equipment is located and having easy access to it is essential should an emergency situation arise so it is important that you always set up equipment according to the specifications of the midwife you are working with

Keeping needed items handy- warm towels, oil and wet washcloths for perineal support, easy access calories and fluids for mom are all examples of items which should be kept available to the midwife you're working with. Again, it is appropriate to ask her ahead of time what she likes to have close by and where to find those items so you're able to anticipate her needs and keep those items nearby.

<u>Notes on gathering information</u>-You will likely be asked to collect information such as:
- Mother's vital signs
- Fetal Heart Tones (FHT's)
- Progress of labor
- Notable events & their respective time (onset of labor, complete dilation, crowning, delivery of the head, delivery of the placenta, etc.)
- Ins & Outs- monitoring everything going into laboring mother's body and everything coming out
  - Example: Water intake-1:30am
  - Bowel movement- 2:20am

Limited Access Birth Assisting (Community Birth Responder)

When assisting in a limited access environment where equipment, clean water, and even safety are not a given, it's important to gather information which may be pertinent to the midwife you're assisting.

For example, if assisting in a remote village birth center your ability to assess the situation and assist a midwife could mean the difference between life and death. Whether you have access to the most advanced medical equipment in the area or are left without any supplies for birth, you must remember that babies know how to be born. <u>Every supply we use in birth in the West is a tool but not a necessity</u>. Even cord clamps and sterile scissor are not necessary if unavailable as the cord can in fact be left in tact (attached to the placenta) until it falls off on it's own and baby will not be exposed to any additional infection risk.

Keys in Limited Access Scenarios:

- If available, ask for help! If the midwife you're assisting has delivered in this environment before, ask her how to best serve her during the birth.
- Gather supplies. Birth tools don't need to be wrapped in sterile packaging to be helpful. If you have stainless steel tools at your disposal, boiling them for 20-30

minutes at a rolling boil will kill all bacteria and viruses and they will be safe to use for a new client. Gather clean towels or blankets to dry and warm baby and use what's available to you.

- Invite the Lord! Your intercession is powerful. Pray for the laboring mom and her baby along with the birth team you're working with. Speak words of life over the baby and don't underestimate the authority your words carry in every circumstance!

Notes:_____

Birth is the sudden opening of a window, through which you look out upon a stupendous prospect. For what has happened? A miracle.
You have exchanged nothing for the possibility of everything.
William MacNeile Dixon

The Scope of a Birth Assistant

So what is the scope of a birth assistance? First, an assistant is not a midwife nor is she a doula. Your scope will be much greater than that of a doula, although many birth assistants do function as doulas too within the birth setting. A birth assistant is an assistant to the midwife, and in that the full responsibility of the laboring mother and soon to be newborn is not hers to carry, but she very much carries it none the less. Most, but not all, birth assistants are student midwives, or plan to be student midwives and will one day maintain practices of your own. Assisting in births will give you the feel for and understanding of birth and its many variables. It is very important to understand that as birth assistants you will carry the majority of the practical and physical needs of birth, and the midwife will carry the diagnostic and ultimate responsibility, along with the mother, for how the birth proceeds.

On a personal note, being a BA was the hardest job I ever held, and yet the most freeing. Being a birth assistant allows you the opportunity to learn and apply your knowledge while still under the supervision of a certified midwife. There is safety in this setting, and I would encourage you to embrace the role of BA, work hard, commit 110%, absorb every detail and make the most of every opportunity.

Birth assistants are trained professionals and should be competent and capable within side their scope or range of practice. In every birth, it is the responsibility of the birth assistant to know the expectations of the midwife she is serving alongside and meet, or exceed, those expectations without continual reminders. The birth assistant should be independently minded enough to understand their scope of practice and responsibilities. It should never be the primary midwives role to remind you of the role you fill inside the birth room. If vital signs and FHT are within your scope of practice then it is your responsibility know the protocols of your midwife and follow them without asking. For example, if FHT are to attained every hour within the early stages of labor then you as the assistant should initiate the gathering and recording of this data every hour on the hour or in the manner of the primary midwifes protocol.

For our purposes the scope of a BA consists of, but is not limited to, the following:

Set Up/Clean Up

This includes:

Setting up the tub prior to birth

Setting up the equipment prior to birth

Preparing the herbal peri wash and PP pads

Wiping down all equipment after the birth with Clorox wipes

Emptying the Tub

Cleaning, deflating and packing the tub away

Laundry

Trash removal

Gathering pertinent information such as:

Mother's vital signs

Fetal Heart Tones (FHT's)

Progress of labor

Notable events & their respective time (onset of labor, complete dilation, crowning, delivery of the head, delivery of the placenta, etc.)

Ins & Outs- monitoring everything going into laboring mother's body and everything coming out

Example:

Water intake-1:30pm or 1330

Bowel movement- 2:20pm or 1430

Medications: time. Location and amount given, expiration date and LOT #

Gathering this information is the job of the assistant and should be gathered, according to protocol, without the midwife asking each time for you to gather this information. For example, if protocol states FHT are taken every 15 minutes during transition, then you as the assistant need to note the time, gather and record this information every 15 minutes under the supervision of the midwife.

Communicating Information

As the assistant it is your job to gather the information, but it is the midwives job to determine what the information indicates and how best to proceed with the information you have gathered. When gathering FHT the midwife will likely be in the room and can also assess the FHT at the same time you gather them. However, when vital signs are taken be sure to communicate to the midwife when you have completed the assessment and your findings.

Keeping Records/Noting Times

Often in birth there is so much going on it is easy for forget times or sequences of events. As the assistant this is one of the most basic, and yet foundational jobs you will have. Take proper notes. If it happens, write it down. I always suggest a running log in a notebook where you can later transfer these details to the formal paperwork. It is sometimes hard to be neat and fast at the same time, if you keep a notebook you will be able to transfer the information to the chart in a more accurate and neat way, without mistakes. During labor and birth the most important part is that you obtain the information. Make sure you are familiar with the charting system the midwife uses. When noting times be sure to always note the following:

any position change, food she takes in, any waste that comes out, especially when she empties her bladder, when she gets in the tub or when she gets out, when any medications or medical oxygen is given or stopped, when she moves from early to active and from active to transition, note the time when she is "pushy" and when she is actually pushing, note the time of visualization of head, crowning, birth of head, birth of body, birth of placenta, time cord is cut, and time the baby has the first protein feed.

It is important here to note the use of the 24 hour clock, also known as military time, in the midwifery field. If you are not familiar with said way of keeping time, this is the occasion to learn. The United States is one of the only countries where the 12 hour clock is used, with English speaking Canada being the other. **All times recorded in a birth should be recorded under the 24 hour clock.**

24 Hour Clock

1:00 am	0100	1:00 pm	1300
2:00 am	0200	2:00 pm	1400
3:00 am	0300	3:00 pm	1500
4:00 am	0400	4:00 pm	1600
5:00 am	0500	5:00 pm	1700
6:00 am	0600	6:00 pm	1800
7:00 am	0700	7:00 pm	1900
8:00 am	0800	8:00 pm	2000
9:00 am	0900	9:00 pm	2100
10:00 am	1000	10:00 pm	2200
11:00 am	1100	11:00 pm	2300
12:00 am	1200	12:00 pm	2400

The greatest joy is to become a mother;
the second greatest joy is to become a midwife.
Norwegian Proverb

What is a Midwife

A midwife is trained specifically as a care provider for mothers and newborns during the pregnancy and postpartum period. Midwives provide traditional methods of support and care and rely on the science of health prevention rather than medical technology. The hallmark of midwifery care is based on individualized and relational care of the pregnant woman. Midwives treat women as a whole and work not only to maintain the health of their body, but also their mind, emotions and spirit during the childbearing years. The term midwife means to be with woman, and this relational aspect of care is what sets a midwife a part from any other care provider. Internationally a midwife is defined as one who has completed an endorsed educational program within the country where they choose to work and serve. Midwives are recognized inmost regions as an autonomous practitioner and is vital to the role of women globally in their childbearing years. Midwives are educators above all and work in partnership with the women and peripheral medical providers to achieve their highest possible standard of health.

What is the Scope of a Midwife

According to the International Confederation of Midwives Global Standards for Midwifery Education a midwife should be competent in the following:

Reproductive Health
Sexual Health
Pregnancy
Labor

Postpartum

Newborn Care

Family Counseling

Childbirth and Parental Education

Woman's Advocacy

Health Promotion

Disease Prevention

The Routes to Midwifery

Certified Professional Midwife

A CPM is a Certified Professional Midwife. CPMs are credentialed through the North American Registry of Midwives (NARM) and are recognized as autonomous birth professionals in 29 states within the United States as well as a few developing nations. CPMs are non-nurse midwives and generally work in out-of-hospital settings such as free standing birth centers, but more commonly CPMs attend births in the home. As a CPM candidate you must apprentice under an approved preceptor for a minimum of 2 years and pass a nationally recognized written and clinical exam prior to certification. To maintain certification, midwives must be in good standing with NARM and maintain current CPR and NRP certifications along with continuing education courses. Certified Professional Midwives are also now recognized by the American College of Obstetricians and Gynecologists (ACOG) as credible and appropriate practitioners in the care of healthy pregnant women.

Certified Nurse Midwife

A CNM offers a full range of healthcare services for women and this includes most primary care needs such as: gynecological checkups, family planning services, preconception care, prenatal, birth, postpartum care and well-baby care. CNM's are nurses first and are recognized in all 50 states and most developing nations. CNM's generally do not practice autonomously, but rather practice under the direction of a physician. CNM's most commonly practice inside a hospital setting.

Certified Mercy Midwife/Certified Community Midwife

A CMM/CCM provides a host of services overseas. These are midwives that do not require a nursing degree to practice but must go through a didactic training regimen as well as an apprenticeship that in total may take 1-3 years to complete. She works predominantly in developing communities that are often rural in nature. At times she may be the only trained health care provider that the community has access to. As such, she will not only provide the suite if typical midwifery services such as prenatal care, labor, birth, postpartum, and well woman care but will also work as a general health care worker in the vicinity. She might be treating cases of malaria, suturing cuts from accidents, and any host of other things that she finds needs attention from day to day. These women are often the only help a community has as other clinics and hospitals are too far to reach.

Certified Midwife/Licensed Midwife

The CM and LM designations are given to a group of midwives that have received their certifications in slightly different ways. The CM designation is a masters program offered in only a select few locations. There is no requirement to be a nurse beforehand for this certification. The CM is limited to only working in a few states within the US though the degree is managed and awarded by the ACNM which also certifies nurse midwives. The LM certification is given by some states in the US after a host of training requirements is met and a qualifying exam is passed. The exam is usually the NARM exam required if CPMs. Both the CM and LM are limited to a few US states and will work in either home birth or birth centers depending on the allowances of the state they work in.

The midwives, however, feared God..
Exodus 1:17

The Spirits of Midwifery

Midwifery is spiritual. Often, midwifery embraces every spirit but the holiest one. It is in our day the Lord is reestablishing his standard and raising up a generation of women, of midwives, by which to restore the picture of birth. For when birth is restored, as God intended, the eyes of man are removed from themselves and placed on the Father.

...and the Spirit of God was hovering...
Genesis 1:2

There are several familial spirits, ones by which midwives have often been marked. Midwives who will choose to walk in the opposite spirit will change nations, and it begins with us.

The Spirits of Midwifery	The Opposite Spirit
Humanism/God Complex	Humility
Witchcraft	The Holy Spirit at Work in Us
Fear	Perfect Love/Brave/Fearless
Bitterness	Sweetness/Fragrance
Rebellion	Submissiveness
Self	Abiding in Him/Co-Laboring
Empowering	Victorious
Divisiveness	Unity

©2016 GoMidwife

The Culture of Death

When we step in front of death and say, "No" it is inevitable we will be noticed. Often when one thinks of midwives they have this picture of someone playing with sweet wrinkly newborns, and we do, but birth is dirty; it is trench work, especially in the spiritual sense. Midwives usher in life, support life, encourage life. We intercede for life and that flies directly in the face of the pervasive culture of our day. It is a beautiful thing to be sure, but it is a serious endeavor as well. We are dealing with some of the most important events in a woman's life and in the growth of the family as a whole, giving us a strong and powerful responsibility. We will have opposition. However we are not to fear anything, but the Lord. We must choose to be a culture of life in a time of death.

Shiphrah and Puah

The culture of death is not new, it is the same culture the Hebrew midwives knew in Egypt. According to Rabbinical tradition, the midwives were mother and daughter which means one generation taught the next; the original apprenticeship model. More poignant than simply being a mother and daughter is the belief that considers the mother and daughter to be Jochebed and Miriam. Their righteousness is celebrated, because they did not fear man, but rather they loved God. They found blessing and favor before the Lord because they chose to abide in Him, they chose to be a culture of life in a time of a death. It is in our abiding in his presence where we find safety, even in the darkest of places or the scariest moments. No one says being a midwife is safe, in fact for his sake we face death all day long[1] that we might cultivate the opposite spirit which is life. The Lord went on to honor the midwives for their fearlessness, by establishing families for them. So we, as midwives who move in opposition to the culture of death are hidden, as Jochebed hid her own child, and what we birth and what we bring bring forth is kept by the Lord. If this holds true then part of the family the Lord

---

[1] Romans 8:36

established for the midwives was a son, named Moses who would lead his people out of slavery, and the midwife who refused to bow to Pharaoh gave birth to freedom.

What does that say for midwives in our day? It says if we will abide in Him he will not forget us. Though we are but handmaidens who kneel before those we serve if abide in Him, co-labor with him, set our hands to establish a culture of life, we will thwart the plans of the enemy as did Shiphrah, as did Puah and we will lead our people to freedom. Indeed, we will usher our people into the Land Long Promised.

Notes

The single biggest problem in communication is the illusion it has taken place.

George Bernard Shaw

Communication

Effective communication will make or break your relationships as a professional birth worker. The ability to clearly communicate expectations and information is essential to positive relationships and your role as a great birth assistant and midwife depend on it.

Communicating with the Midwife

Communication with the midwife you're working with is not limited to your time together at a birth. It's important that you discuss the expectations you both have for your professional relationship and that she knows she can count on you when she needs you most. Some important topics to cover before you ever attend a birth together might include:

Availability: Your role as a primary birth assistant may require flexibility and being on call is very often part of the job description. Communicating your availability in advance will help appropriate expectations to be set and will help you to be able to plan accordingly.

Your expected role in an emergency: Your training as a birth assistant has prepared you for the unlikely event of a birth emergency and it is important that you know what role you would be expected to take on should an emergency arise. An example of emergency role would be taking responsibility for effective neonatal resuscitation using positive pressure ventilation (PPV). It should be determined ahead of time where your strengths as a birth assistant lie and where you might require some more training in order to feel secure assisting in specific areas.

Your communication style during labor is especially important. Consider the following as you determine positive ways to communicate during birth.

Important details throughout labor: The role of a birth assistant requires consistent observation and conveyance of pertinent information to the midwife. You will serve as her second pair of eyes and should report any information that will help her to serve the laboring mother effectively. Be sure to speak calmly and take care to share what's important for the midwife to know without bringing fear into the scenario.

The timeliness and delivery of this information must also remain professional. In being asked to monitor a client, the midwife you are working with will expect that you know when the information you're receiving is outside of normal limits. For example, if a mother's heart rate has fallen well below her baseline and is remaining there, it's important to share this information with the midwife quickly. Understanding the basics and being willing to ask for clarification when you need help are a big part of being a great birth assistant.

Maintaining a professional and peaceful demeanor is especially important in an emergency situation. Remember to stay calm, take a second to collect yourself and think through what you know in any given situation. If your role isn't clear and you have the opportunity to communicate with the midwife without distracting her from her current task, ask how you can be helpful. However, keep in mind that if the midwife is

Take care to communicate with the Lord throughout the birth, praying for the mother, baby, their family and everyone present at the birth. This communication with God will help you maintain a peaceful demeanor and your peace will actually help to keep the situation free of anxiety.

Communicating with Clients

Client communication should always reflect the utmost professionalism, whether you know the client personally or not. This is not to say that you should not speak kindly and personally. However, professional communication requires a level of trust and confidentiality that she needs to feel between you in order to ensure comfort during her birth.

When communicating information to a client, remember that her pregnancy is personal to her. All information should be conveyed kindly and respectfully, taking into account that knowledge of medical terms is not always a given. Take time to explain things when you have understanding and be willing to ask for help when you do not. Always comply with HIPPA regulations in order to protect your client and her privacy. Never discuss her personal information with anyone outside of a peer review format and never without her permission.

Communicating with Laboring Mothers

Communicating with a laboring mother is an extra sensitive job. Your words convey much but your body language, overall demeanor and emotional state are also communicating when words are not. When communicating with a woman in labor, look her in the eye and speak softly but clearly. Explain what needs to be explained in as few words as possible, remembering that if she needs you to elaborate she can always ask you to do so. It's important that we keep communication simple and concise as mom is likely working to keep herself free of distraction in order to complete the task at hand.

Communicating with Peers

Professional communication with peers is an incredibly effective way to gain a respected role in the birth community. A culture of honor promotes the fruit of the spirit and you hold the power of life and death in your words! Remember that anyone seeking to serve women and their children safely is on the same team you are, regardless of differences in values, religious beliefs/preferences or protocol. Relationships can be damaged so easily by flippant talk so it is important to decide in your heart that you will speak words of life over those you serve alongside and establish a gossip-free position. There is a temptation among birth workers to allow competition to drive women to slanderous talk about one another but this must not be allowed to continue! The determination of one birth assistant to stop this competitive gossip will lead to further life giving speech and makes room for the Lord to be active in our practice. Remember to bless those you work alongside as well as those with whom you share not only a professional similarity but a heart to see women empowered and loved well as they bring forth life of their own.

Notes_____

To be a professional you have to act like one as well.
Alcurtis Turner

Professionalism

As a birth worker, you should seek to be professional and represent yourself well! We've discussed the importance of communicating well and of considering the impact our words and demeanor have on the level of respect you will receive but there is more to professionalism than communication.

Look Professional

Representing yourself as a birth professional starts with the way you look. This is not to say that you should consider your outward appearance more important than say, how you speak to clients, but remember as an assistant and as a student midwife, you represent the midwife you are working for or with, and it is vital you represent yourself well, so she too is represented well. The way you represent yourself physically does communicate more than you may think. A messy appearance portrays the image that you don't care about your appearance or couldn't be bothered to put effort into your appearance.

Some rules of thumb for professional appearance:

<u>Consider your audience</u>: If you are working in communities or among cultures where women dress more conservatively, do so yourself. Being willing to cover up is a matter of respect and shows that you value the individual you're serving. This is especially true among Amish or Mennonite communities in rural areas of the US and in traditional communities around the world.

<u>Make hygiene a priority</u>: If showering regularly isn't part of your routine, now is the time to develop a hygienic regimen. Frankly, body odors and foul breath are especially noticeable in the close proximity that birth-assisting will often require. Showering regularly and paying mind to your personal odors and appearance are important parts of your job. It is especially important during the birthing process. Often births are long and you will stay in the home for

hours, if not days. Always pack yourself a toiletries kit in order to be able to maintain your appearance. In the event of a transfer you will also want to freshen up prior to accompanying the mother to the hospital. If there has been an emergency, you do not want to arrive to the hospital wild and array.

<u>Put in the effort</u>: Keeping nails trimmed and clean, making sure your hair stays out of your face, wearing clothes that allow you to move freely and serve without distraction will not only make your life easier while you work but again, helps convey a level of professionalism that's important in your job as a birth assistant.

<u>Go with the flow</u>: Ask the midwife you're working with if she has a suggested dress code for interviews with potential clients or for the birth itself. She may require that you wear semi-professional dress for interviews and scrubs during births or she may not have a preference whatsoever.

Representing Your Midwife

Regardless of what you're wearing, be sure that you are representing the midwife and her birth team well until you become the midwife yourself. Your attitude, willingness to learn and accept constructive criticism and your desire to serve all reflect on the midwife with whom you are working. Ensure that you conduct yourself in a way that the midwife will be proud to include you on her birth team and that reflects the value you feel for your work and for the families you are serving.

Anatomy

Ovaries

- Located within the broad uterine ligament on either side of the uterus, below the fallopian tubes
- The ovaries house and release the ova (eggs)
- At birth, each female has anywhere between 1 and 2 million eggs but less than 300 will be fully formed and released for reproduction
- Upon release from the ovary, the ovum (egg) travels down the fallopian tube attached to the releasing ovary toward the uterus for fertilization and eventually, implantation.
- The fallopian tube is a muscle which connects the ovary with the uterus.
- Tiny hair-like structures called cilia line the interior walls of the tube, which help push the ovum down toward the uterus after it is released from the ovary.
- Average length is 10cm (4 inches)

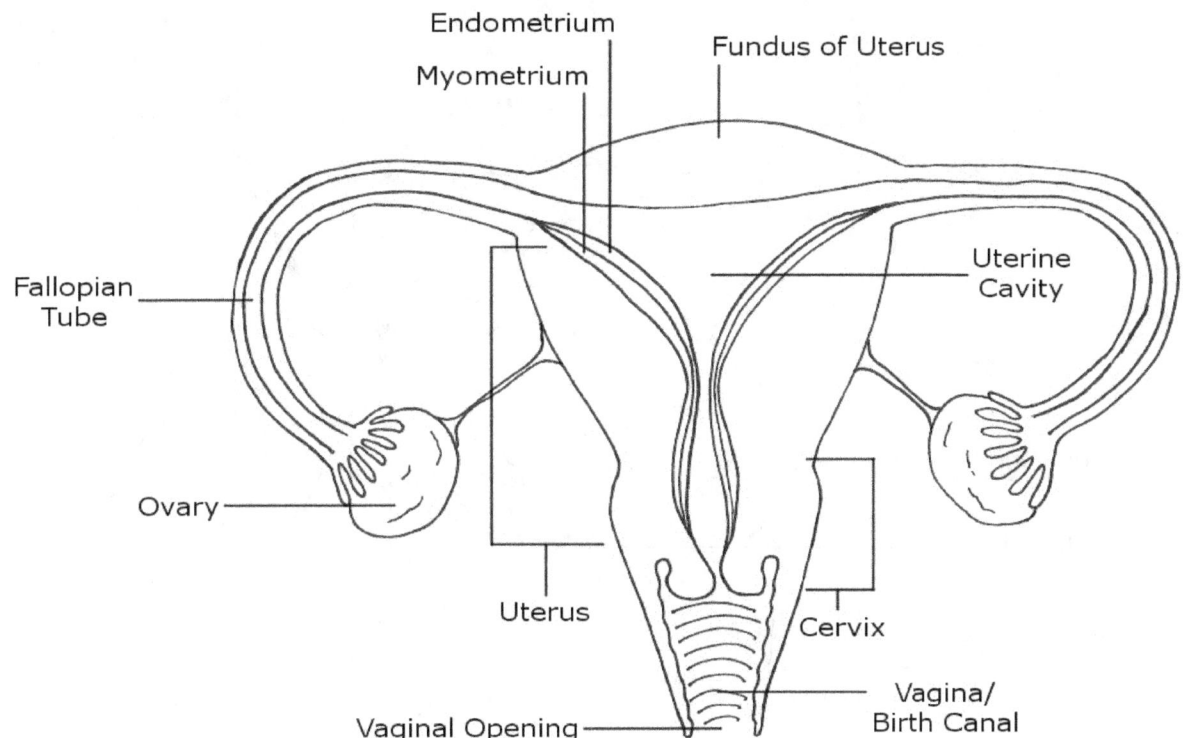

© Copyright 2015, GoMidwife

Uterus

- The "normal" uterus resembles an upside down pear- usually between 3-4 inches long by 2-3 inches wide when not pregnant.
- Main structures of the uterus:
  - Body of the uterus: implantation site of a fertilized ovum
  - Cervix: lowest portion of uterus at top of vagina
  - Endometrium: uterine lining
  - Myometrium: muscular wall of uterus
  - Fundus: top (rounded) portion of uterus between the two fallopian tubes

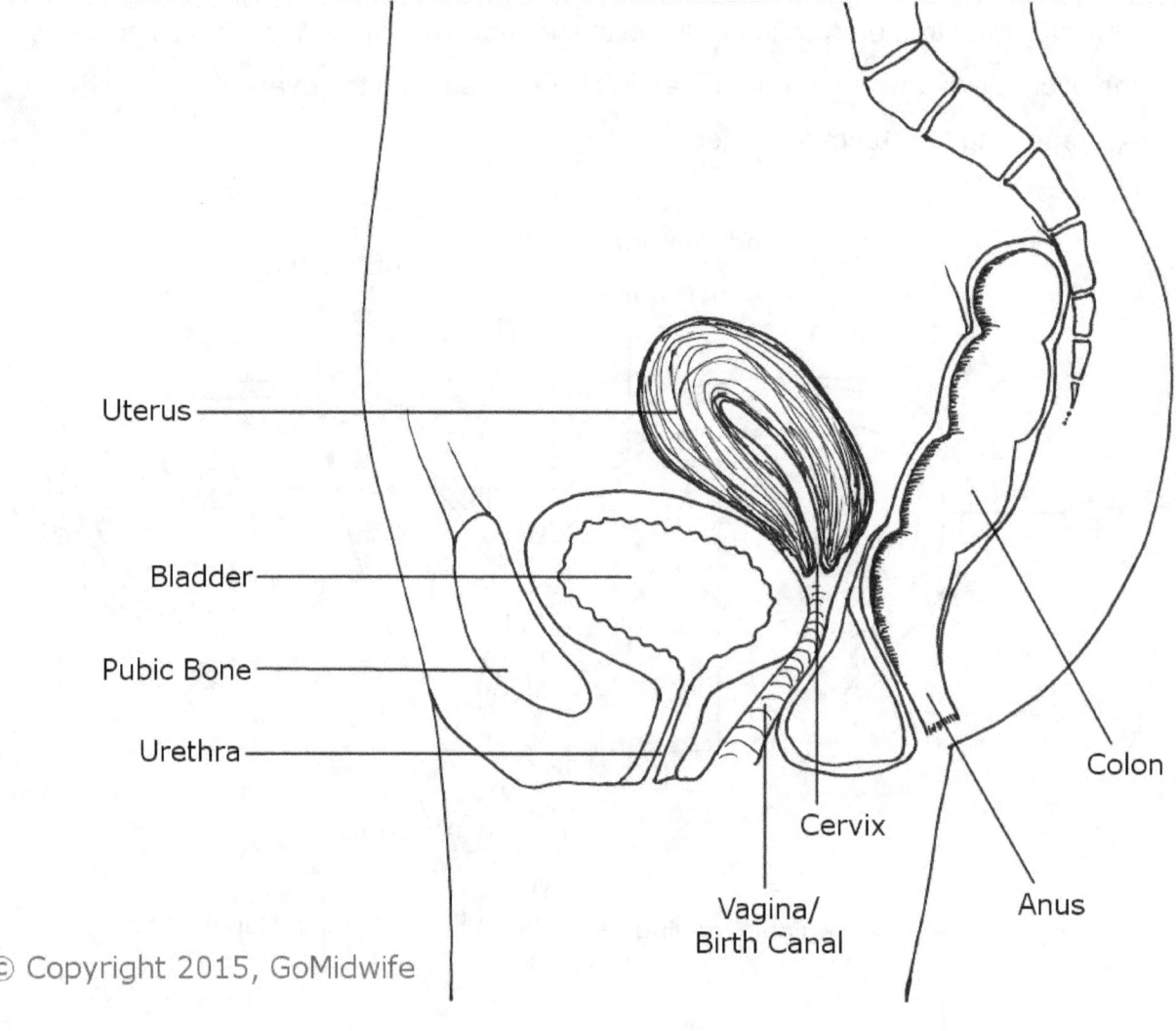

Cervix of the uterus
- The cervix is the lower portion of the uterus which leads to the vagina (birth canal)
- During a menstrual period, blood flows out of the uterus through the cervix.
- The cervix makes mucus which help sperm move through the vagina into the uterus in order to fertilize an egg & create life!

Cervical effacement and dilation during birth
- During pregnancy, the cervix stays tightly closed in order to protect baby and to keep him inside.
- As labor begins, the cervix will thin and retract up toward the top of the uterus- this is called effacement.

   Effacement
   - Effacement is the term used to describe the thinning and ripening of the cervix for birth.
   - Effacement is measured in percentages (25% effaced, 100% effaced) and the percentage is meant to gauge how much the cervix has softened and begun to recede into the uterine wall to make room for the baby to be delivered

   Dilation
   - Dilation of the cervix occurs during the laboring process. During uterine contractions, the cervix will open wider until it has reached its complete dilation (10cm).
   - It is at this point, when a woman is fully dilated that she is ready to push her baby out.

Vulva
- The vulva contains the outer parts of the female genitals, including the following structures:
   - Labia majora: the outermost layer of tissue surrounding the inner structures of the female reproductive system

- Labia minora: the inner layer of tissue on either side of the vaginal opening
- Clitoris: the external structure housing more than 8,000 nerve endings, thus a highly sensitive female erogenous zone
- Urethral opening: the external structure urine passes through to be eliminated from the body
- Vagina: The vagina is a tubular fibromuscular (both fibrous and muscular) tract that is associated with both sexual intercourse and childbirth
    - Prior to and during intercourse, the vagina produces natural lubrication, which makes intercourse more pleasurable for both parties.
    - During menstruation, blood and mucosal discharge flow from the uterus, through the cervix and into the vagina
    - During childbirth, the vagina is also known as the birth canal and is the last opening through which baby must descend
- Anus: the external structure fecal matter passed through to be eliminated from the body

Anatomy of Birth--- Start to Finish

It all begins in the ovary. The ovum is released at ovulation and carried through the fallopian tube by small finger-like projections which line it's wall, called fimbriae. It is in the fallopian tube that the ovum will meet the sperm sent into the body through the uterus during ejaculation/intercourse. If sperm powerful enough to penetrate the tough wall of the ovum meet the ovum, life begins!

Once fertilized, the zygote (fertilized egg) will continue to move toward the uterus where it will implant into the thickening endometrial layer of the uterus. Here the zygote will grow and become an embryo and finally a fetus.

The fetus will continue to grow and mom will experience practice contractions beginning as early as her 28th week. As mom's body begins to prepare for the imminent delivery of her baby, she will experience a multitude of bodily changes including the shedding of the thick mucus which has kept her womb closed to bacteria and other intruders. As pregnancy progresses toward delivery, baby will engage his head down into mom's pelvis and will be putting pressure on her cervix, or the lowest portion of her uterus. This will cause it to begin to open, or dilate, to make room for baby's exit. When mom begins to experience true and active labor, the uterus will contract at increasingly intense intervals in order to continue to bring baby down and eventually expel baby completely. Baby will pass through mom's vagina, which is called her birth canal only during her birth and out into the world.

Anatomy of the Female Pelvis

Identify each of the pelvic landmarks

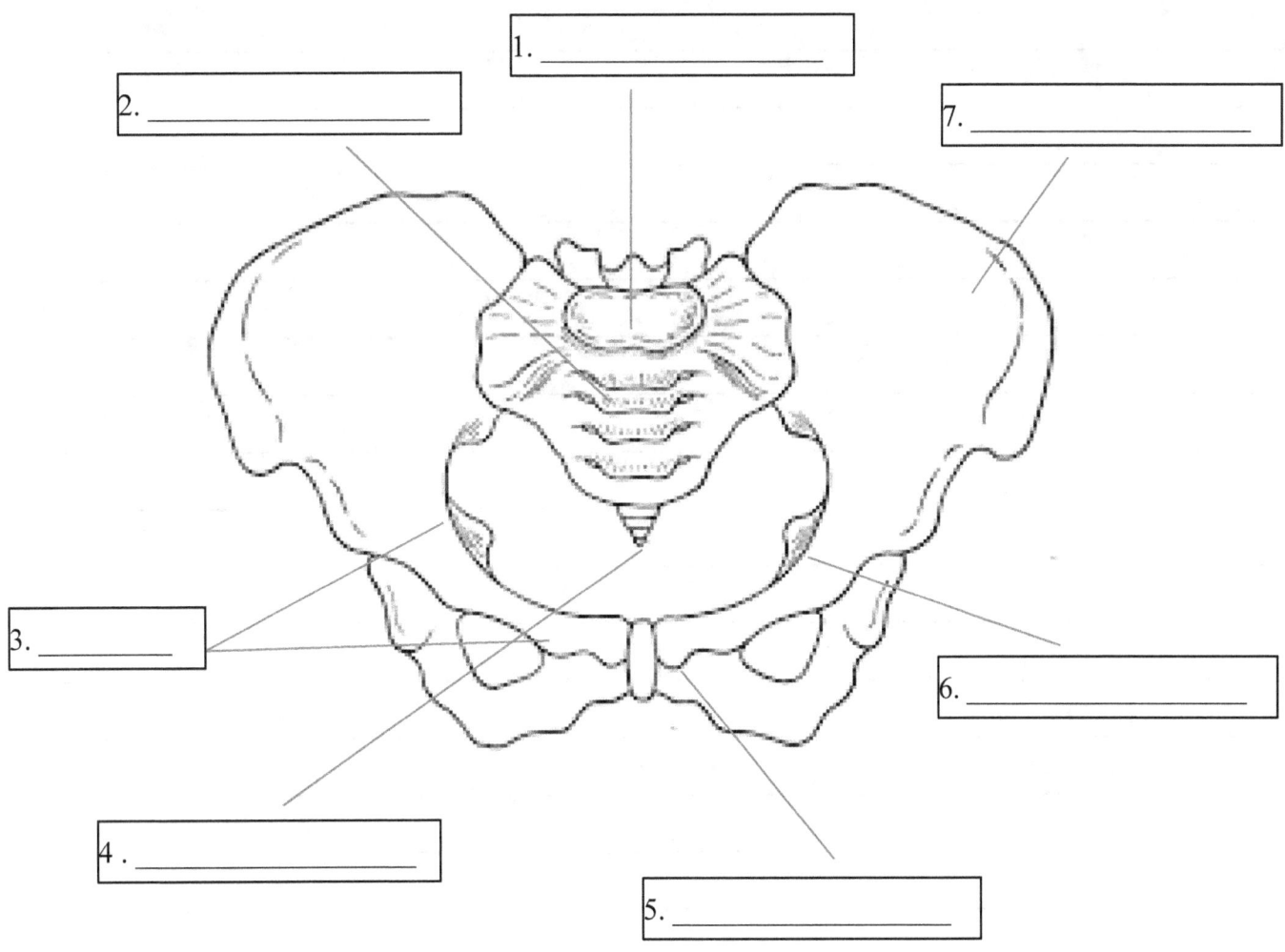

1. _____
2. _____
3. _____
4. _____
5. _____
6. _____
7. _____

Notes

Understanding the Menstrual Cycle

It is surprising really just what women do not know about their bodies and what is actually taking place within in them each and every month. Understanding the menstrual cycle will better equip you to understand conception, hormones and so much better how to educate women during their childbearing years. Even in this, the design is so utterly astounding one must stop and be in awe.

Facts about the menstrual cycle

- <u>Cycle</u>: Counted from the first day of one period to the first day of the next period.
- Average cycle length is 28 days but a healthy cycle can last anywhere from 21-35 days.
- The cycle is marked by three distinct phases:
    - Follicular Phase
    - Ovulation
    - Luteal Phase
- Menstrual fluid is made up of four parts: blood, cervical mucus, vaginal secretions and endometrial tissue

Phases of a menstrual cycle:

Follicular phase

- Begins with menstruation and ends with ovulation.
- The length of this phase varies by circumstance- diet, stress, illness can all effect the length of this phase.

- Follicle Stimulating Hormone (FSH) & Luteinizing Hormone (LH) are produced at the beginning of this phase causing follicles to be released into ovary. Dominant (strongest) ovum will be one released for possible fertilization.

- Estrogen levels rise during this phase which help follicles mature and causes the lining of the uterus to thicken to prepare for potential pregnancy.

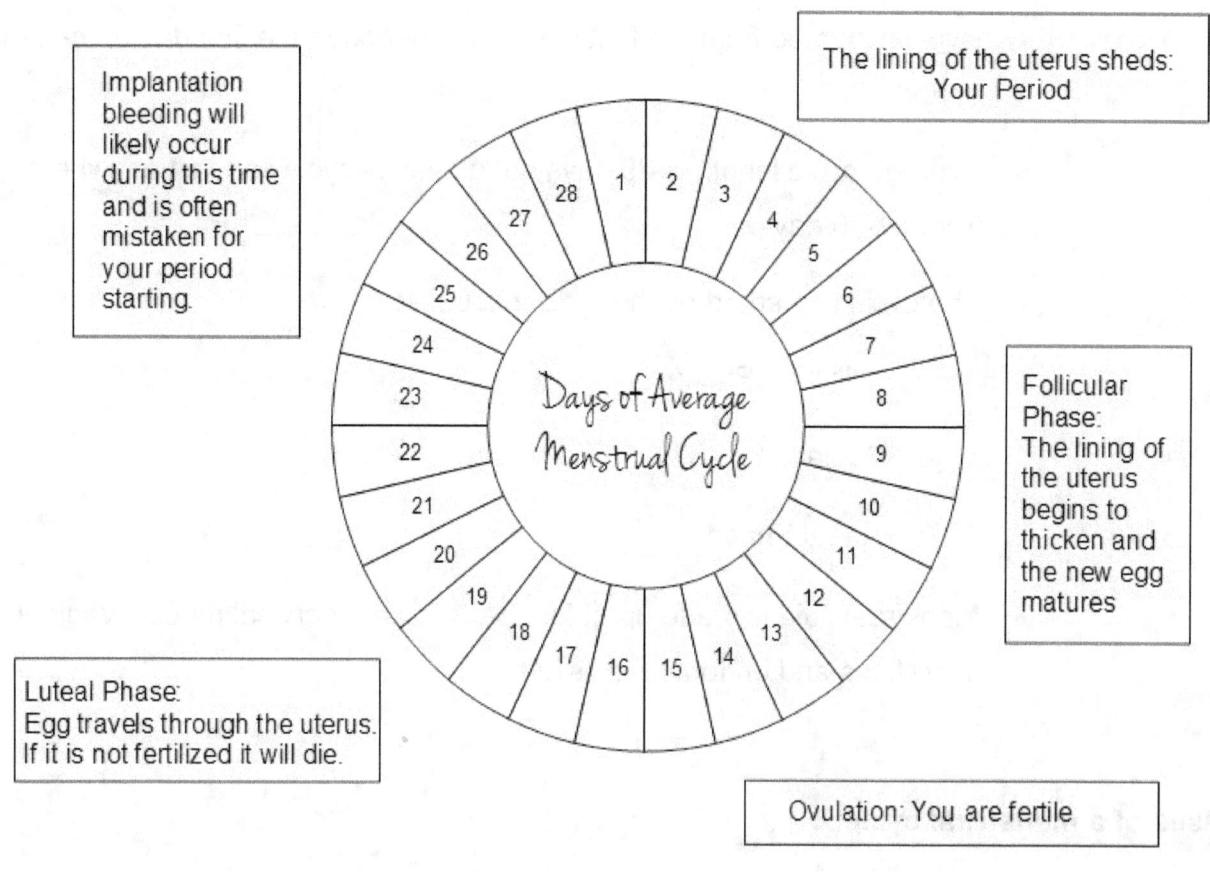

Ovulation phase

- Marked by release of dominant ovum.
- Release of mature ovum is called ovulation.
- Cervix during ovulation:
    - Cervical mucus is generally slippery, clear, and quite stretchy.
    - Mucus help sperm move toward the egg while protecting them from the acidity of the vagina.
    - Feels high, soft and open during this phase.

Luteal Phase

- The luteal phase is marked by an increase in progesterone which is stimulated by the corpus luteum
- This phase does not vary in length by circumstance and lasts 14 days for most women (12-16 days still considered normal)
- During this phase, if fertilization or implantation does not occur, the corpus luteum will shrivel (14 days post ovulation) and progesterone & estrogen levels decrease. This drop produces menstruation.
    - Corpus Luteum: remains of the dominant ovum as it travels from ovary through to uterus

Notes_____

_____
_____
_____
_____
_____
_____
_____

Fetal Development[1]

The development of the fetus is simply put, miraculous. Below you will find an overview of fetal development from conception until birth. Understanding the process and phases of development will help you to better educate women on their nutritional and emotional needs as well be able to help them recognize and understand the milestones within the womb. growth of her baby and wand needs during each stage of growth and change.

Week 1-2

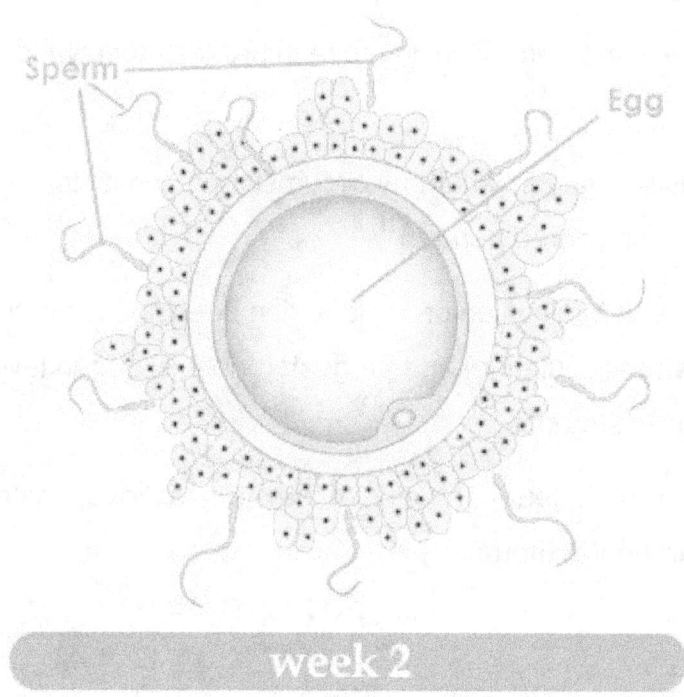

- Normal menstruation occurs
- The hormone that stimulates the release of the ovum (egg) is produced. This hormone is known as FSH (follicle stimulating hormones)
- The ovum, also known as the ovarian follicles, that will be met by sperm are developed
- The ovum is maturing in preparation for release from ovary
- Estrogen begins to be produced

- The hormone which leads to the rupture of the follicle, knows as LH (Luteinizing hormone) is produced
- The ovum (egg) is released into one of the fallopian tubes. This process is known as ovulation
- Ovulation occurs between day 14 and 18

Week 3

- The egg just released from the ovary begins it's journey toward the uterus. These eggs are viable for approximately 12 to 24 hours so the fallopian tubes contain finger-like projections which help move it toward the uterus quickly
- Sperm are viable anywhere from 2 to 5 days and will meet the egg as it journeys toward the uterus if conception is to occur
- If sperm penetrates the egg's outer membrane, fertilization occurs

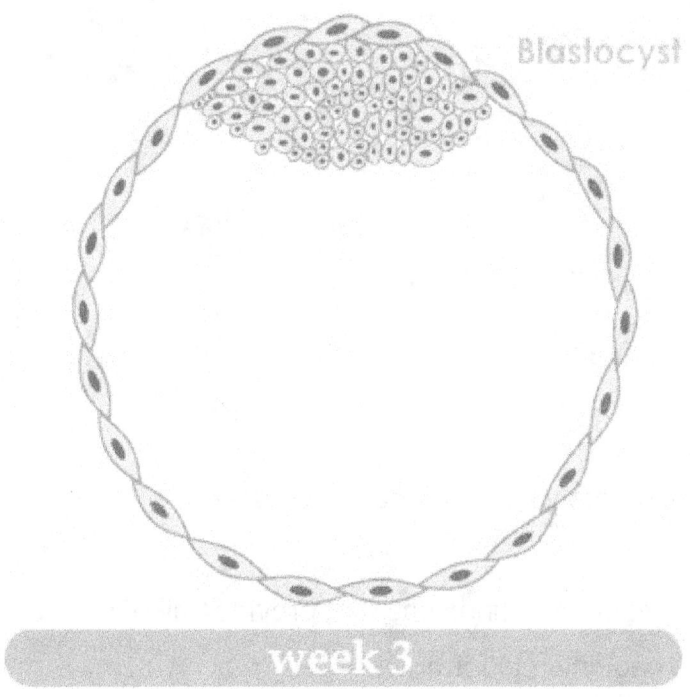

- Fertilization is complete when the nucleus of the egg and sperm have combined

- The gender of the baby is already determined by the time fertilization has completed- baby is now a boy or a girl
- The fertilized egg is now known as a zygote
- The zygote immediately begins to divide and will continue to divide every 12 hours as it descends down the fallopian tube into the uterus to be implanted
- The zygote is now called a morula, which just means "solid ball of cells" and is the result of the zygote's cell division
- The morula will enter the uterus approximately four days after fertilization occurs

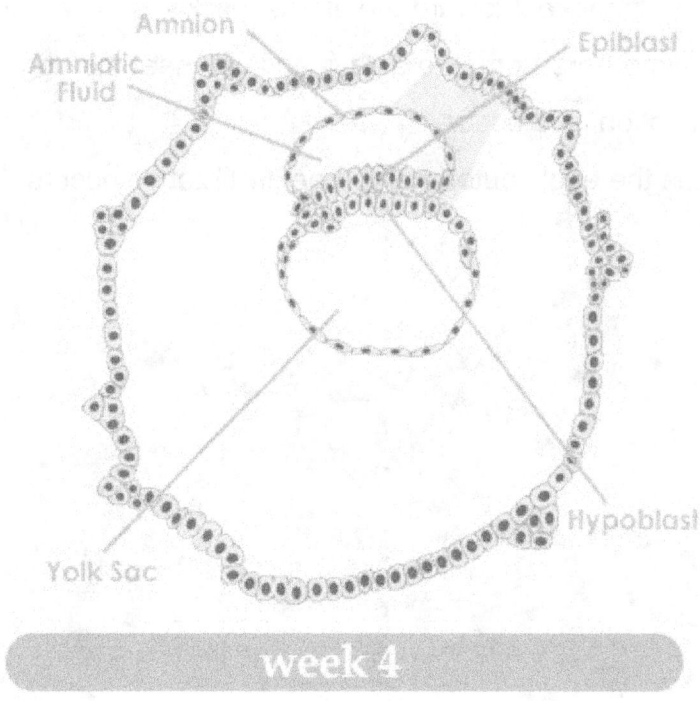

Week 4 -Baby is approximately the size of a poppy seed!

- The yolk sac, a membrane-lined sac that provides all of baby's nourishment in early development is beginning to form
- During this week the morula becomes the blastocyst, which is the mass which will form the embryo

- The egg is beginning to form two groups of cells: the embryo and the placenta
- The blastocyst fastens to the back wall of the uterus and mother and baby are now connected
- Embryonic hormones, or the hormones produced by the baby are beginning to flow through mom's blood so a pregnancy test would show positive during this week.

Week 5- Baby is the size of a sesame seed!

- Mom should be eating lots of foods rich in folic acid (broccoli, dark leafy green veggies and citrus fruits) as they are vital to the development taking place this week
- The brain, heart and spinal cord are beginning to form along with the gastrointestinal tract
- Mom may notice that menstruation has now been missed
- She may experience some spotting and think she is having a normal period but many women will experience some light bleeding as a result of implantation

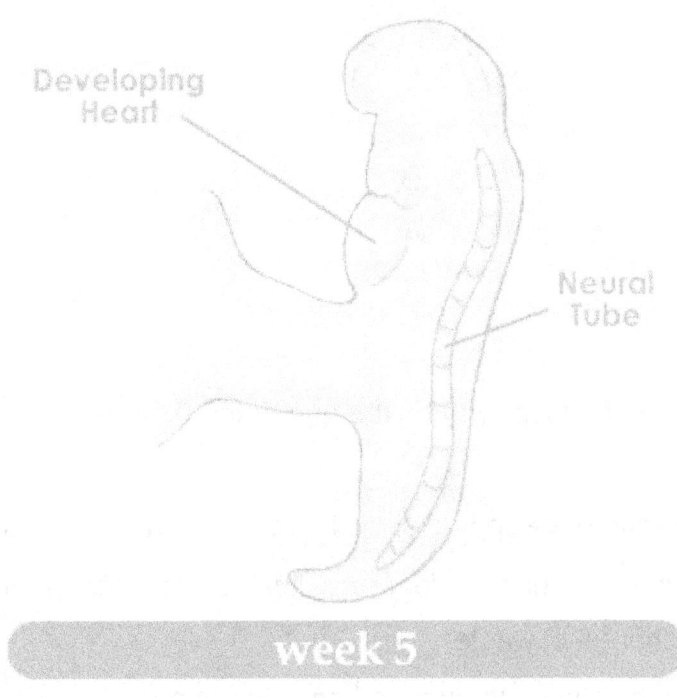

Week 6- Baby is the size of a lentil bean

- Baby's heart will begin to beat this week!
- Arm and leg buds are visible and the structures that will form baby's eyes are forming
- The spinal cord begins to close and blood is flowing through the large blood vessels throughout baby's body
- The umbilical cord has formed and is the size of a thread

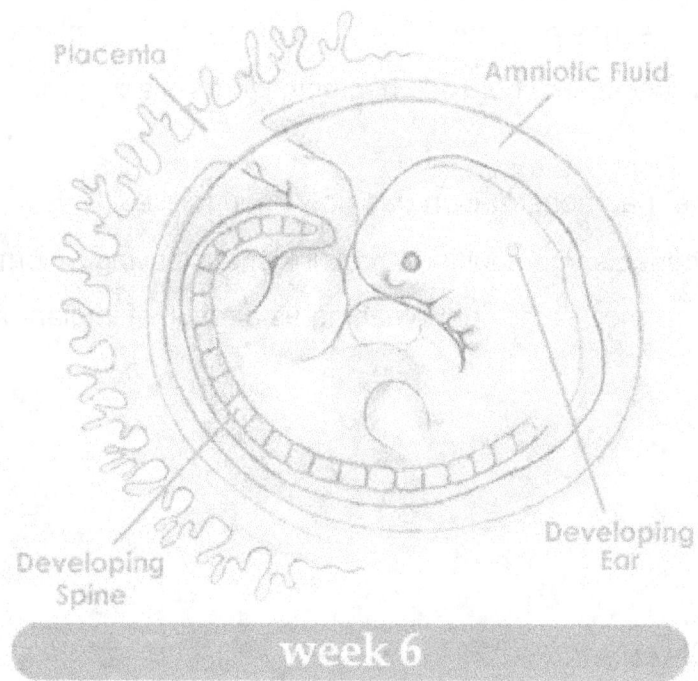

Week 7-9-Baby is the size of a raspberry by the end of the 9th week!

- The brain continues to develop and splits into three parts: fore, mid and hind brain
- Baby is now floating in amniotic fluid, or the fluid which protects him as he grows and helps with the exchange of waste
- The placenta is processing nutrients and passes unneeded waste products to mom through the umbilical cord to be excreted along with her waste. The yolk sac will no

longer provide nutrients for the baby at this point (usually occurring at the end of the 9th week)
- The esophagus is forming along with long buds on either side of it which will later become baby's lungs
- Baby's body is uncurling and stretching out as it grows
- Baby's brain is developing rapidly and brain waves are beginning to form
- Sex organs are developing but won't be recognizable for several more weeks
- Muscles become stronger and baby begins to move, but baby is still too small for mom to feel these movements
- Fingers, toes and eyelids are forming form

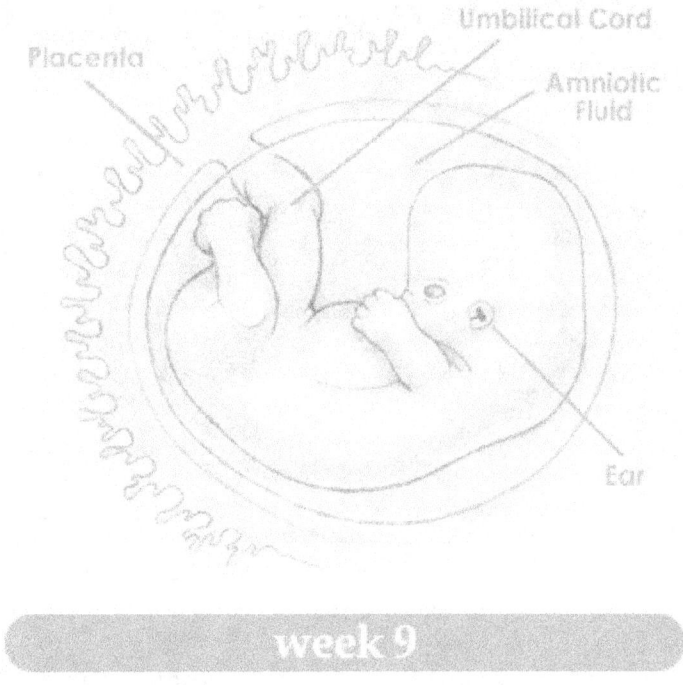

week 9

Weeks 10-12- Baby is approximately the size of a plum by the end of week 12

- Baby's intestines have formed inside the umbilical cord are moving inside baby's abdominal cavity
- At the end of week 10, baby is no longer an embryo. He or she is now called a fetus
- All major internal and external features, organs and limbs are now formed

- Taste buds form on a tiny tongue
- Amniotic sac is filled with about 1.5 oz of fluid
- Tooth buds are forming inside the gums
- Baby's fingers and toes are no longer webbed; fingernails and toenails appear
- Cartilage that shaped the skeleton is solidified and has become bone
- Baby can now swallow, stick out his tongue, get the hiccups and suck his thumb
- Risk of miscarriage decreases dramatically by the end of week twelve

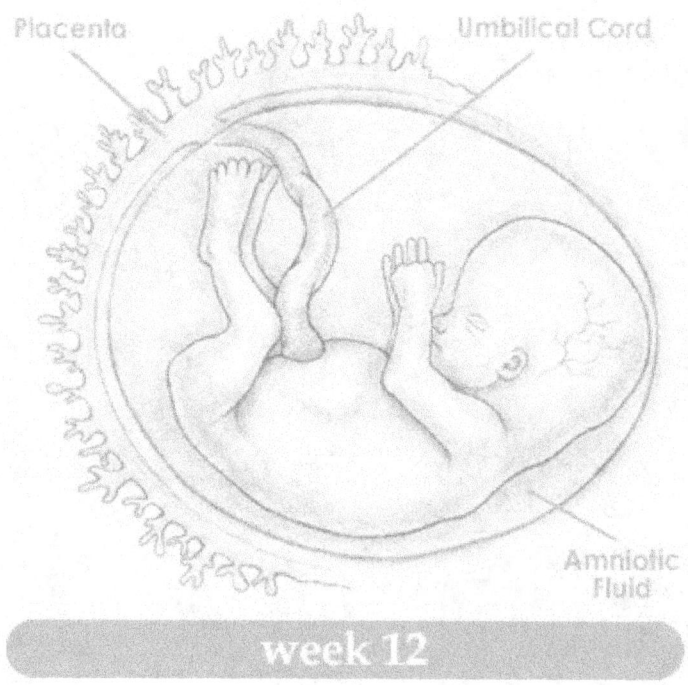

Week 13-16- Welcome to second trimester! Baby is approximately the size of a lemon by the end of the 14th week.

- Mom is 13 weeks pregnant and baby is 11 weeks old
- Baby's eyelids will close and won't reopen until the 28th week
- Her skin is transparent but gaining durability
- Baby will quadruple in weight during this period

- Hair begins to sprout on baby's head and tiny hairs called lanugo will cover baby's body.
- Baby can breathe under water: inhaling and exhaling amniotic fluid which will help the lungs form in later development

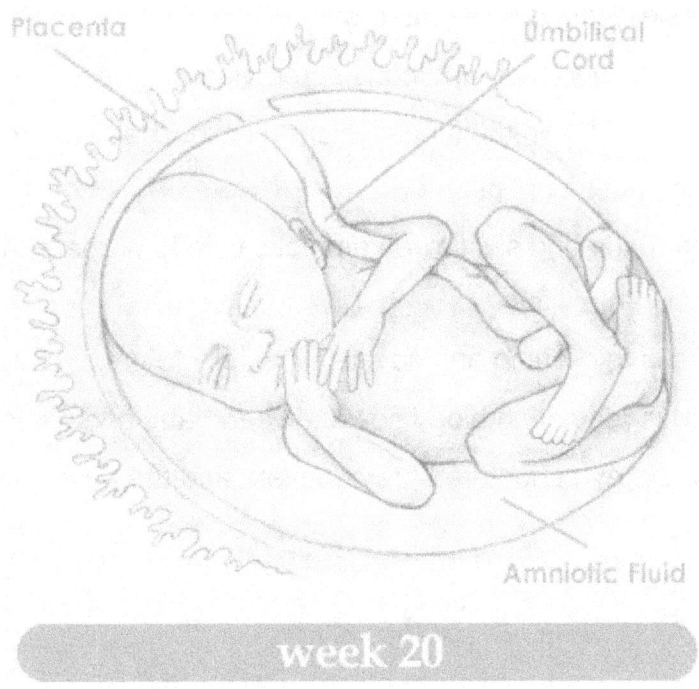

week 20

Weeks 17-20- Baby is the size of a large avocado!

- Mom will begin to feel small movements or flutters in her belly
- Meconium, or baby's first bowel movement made up of cellular waste is developing baby's recently developed intestines
- Baby can hear sounds
- Vernix, or a thick waxy substance which protects baby as skin is developing, begins to cover the body
- Baby's genitals are distinct and visible on ultrasound by the 20th week so parents who want to know their baby's gender are able to find out during this period
- Baby recognizes mother's voice
- Heart can be heard with a stethoscope

Week 21-28- By the end of week 28 baby is approximately the size of an eggplant!

- Baby girls will have all of their eggs by the end of this period- anywhere between 1 and 2 million eggs!
- If baby is a boy, his testes descend and early sperm develop
- Baby has regular sleep and wake patterns
- By week 24 baby is fully formed and considered viable for life
- Surfactant, a fluid secreted in the lungs which will allow baby to breathe oxygen for the first time is produced in the lungs
- Baby's kidneys are excreting swallowed amniotic fluid to prepare them for urine production in extrauterine life
- The umbilical cord is thickening and covered in a jelly to prevent knotting
- The umbilical cord is regulating blood flow between baby and placenta
- During the 28th week, eye structures are complete and the eyes are beginning to open

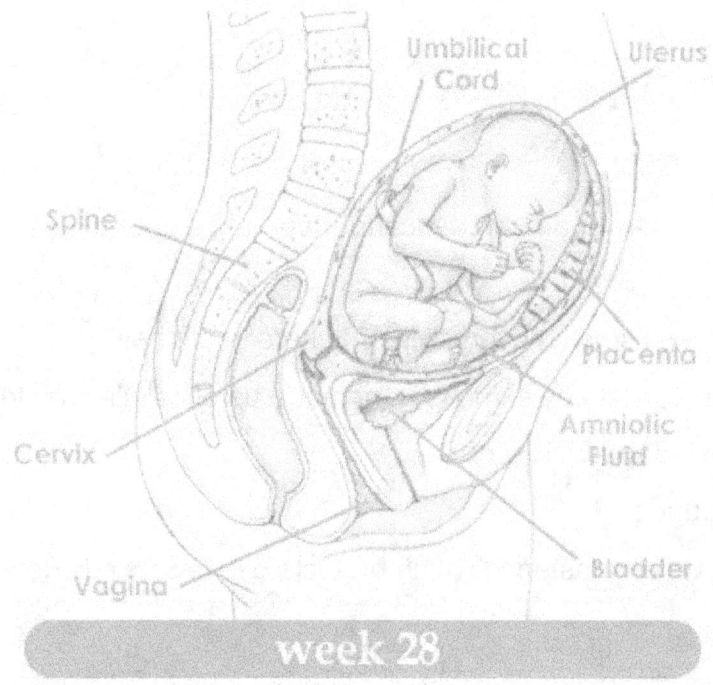

Weeks 29- 34- Welcome to third trimester! By week 30, baby weighs approximately 3 pounds.

- Bones are soft and pliable to allow baby to be born safely but are fully developed

- Baby's brain is developing rapidly and her central nervous system is beginning to control some body functions
- Baby is beginning to gain weight in preparation for birth and rapid fat storage is taking place
- Rhythmic breathing occurs and baby is able to taste and responds to pain
- Baby is beginning to store iron, calcium and other minerals
- She should turn head down in preparation for birth and head will remain engaged in the pelvis

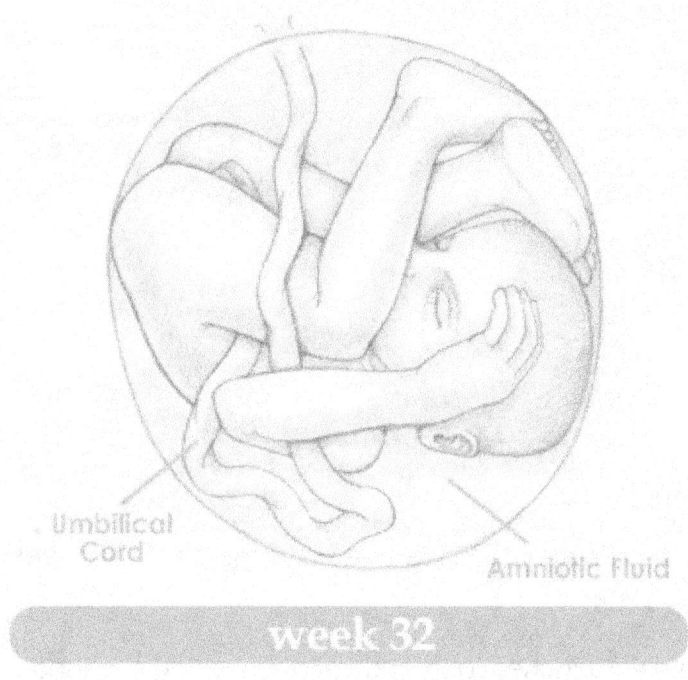

week 32

Weeks 35-38- Baby is approximately the size of a honeydew melon by the 35th week

- Mom will begin to feel practice contractions known as Braxton-Hick's. These contractions are preparing baby as well as mom's uterus for imminent labor
- Baby's body is storing large deposits of brown fat
- Tiny hairs formed during weeks 13-16 (known as lanugo) are beginning to disappear
- Fingernails have grown to reach the end of the fingertips
- By 38 weeks, baby's lungs are fully formed and he can safely live outside of the womb

- Mom may notice a thick, orange or yellow substance known as colostrum beginning to leak from her breasts
- Baby is fully developed and ready to be born

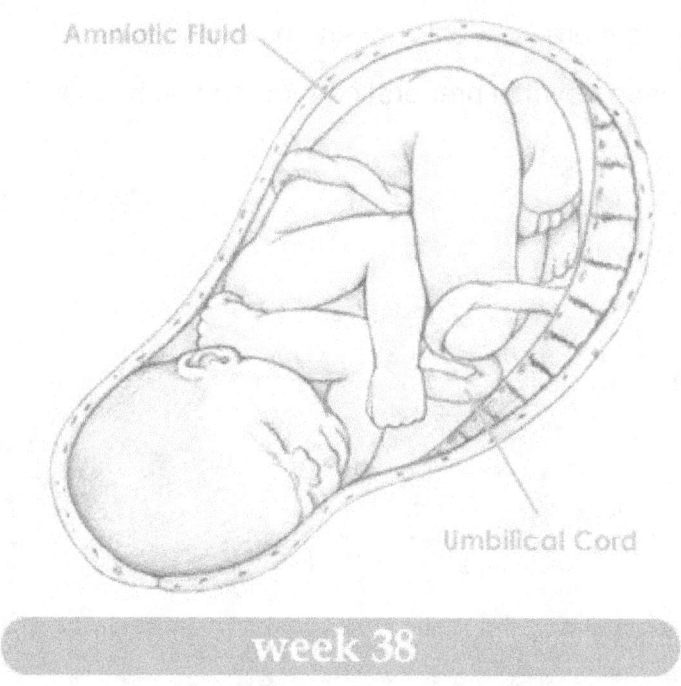

week 38

Weeks 39-42- Baby is approximately the size of a watermelon by 40 weeks

- Baby is approximately 20 inches long and weighs around 8 pounds.
- Her head is engaged in mom's pelvis and she is likely kicking around and putting pressure on the lowest part of mom's uterus, known as the cervix, which will dilate completely to allow baby to pass through. This process of cervical dilation and the contraction of the uterus which accompanies this dilation, along with the decent of the baby through the pelvis, is called labor.

> *Natural childbirth allows the hormones that have been working for women for thousands of years to fulfill their functions. This is more important than just helping a woman through labor and delivery. Birth-related hormones also affect well-being much later in life.*
>
> Janet Schwegel

The Hormones of Conception

FSH

Estrogen

LH

Progesterone

One of the necessary hormones for conception is follicle stimulating hormone or **FSH**. This hormone is present during each cycle and its function is to both produce and release mature eggs ready for fertilization. When the egg is released it is called the follicle and this then produces the next needed hormone: estrogen. **Estrogen** then readies the environment by promoting secretions of cervical mucous increasing the likelihood of conception by improving the environment for the sperm and the fertilization process. Additionally, beyond the environmental preparations, estrogen also releases another hormone called **LH**, or lutenizing hormone which causes the egg to be released from the follicle and the egg is now primed for fertilization. Once the egg is fertilized it must still travel through the reproductive system for six more days until it reaches the womb and implants. This cast off follicle is now known as the corpus luteum. The flow of hormonal production is impeccable and the shed follicle, now the corpus luteum begins the production of progesterone, quite possibly the most important hormone of pregnancy, labor not withstanding. So, what does **progesterone** do? For now, its main function is to support the maintenance of the uterine lining in preparation for implantation.

The Hormones of Pregnancy

hCG

hCS

Progesterone

Relaxin

Estrogen

As you have read, hormones are an intricate part of the process beginning prior to, and in the facilitation of, conception and implantation. Each hormone is a chemical concoction released into the bloodstream in response to stimuli. We have looked at a few hormones which comprise much of the conception process, now it is essential to understand the hormones involved throughout pregnancy, their purpose, function, and the resulting emotional and physical responses women may encounter along the way. As we learned, the progress of hormonal production is a sophisticated design and rarely in need of our outside involvement. Ever more paramount it is to support, and not disrupt, inhibit or induce the natural flow and sequence of hormones in the female body. These are multiple hormones that sustain pregnancy, these are the ones you should know:

Human Chorionic Gonadotropin hCG- almost all of the hormones for pregnancy are produced by the placenta and hcCG is no exception. It is one of the first hormones to present, generally around day six, when the fertilized egg implants into the womb. When this occurs hCG is released signaling the new mother she has life inside. Once these hormones are released they travel through the maternal bloodstream and her body recognizes and begins to adapt to the pregnancy. During this time, a message is sent to the pituitary gland and her menstrual cycle will cease. Levels of hCG grow exponentially during pregnancy. Commonly at 3 weeks you will see levels near 5-50 mIU/mL then you will see a rise to 1,080-56,500 mIU/mL at 6 weeks, reaching peak numbers around week 12 then lowering to a sustained production resulting in levels of hCG up to 117,000mIU/mL in the third trimester. The increased levels of hCG are thought to be a cause for nausea and vomiting in early pregnancy. HCG is eliminated through the urine and can be detected as early as 7-9 days from implantation. HCG stimulates blood flow to the female pelvic region as well to the kidneys. This increase in the hCG hormone and the subsequent stimulation of blood flow is considered the main cause of increased urination during the first trimester. HCG is flushed out through the urine and this hormone present in the urine is what indicates pregnancy on a over the counter pregnancy test. **Human Chorionic Somatolactropin hCS** also known as **hPL or human placental lactogen** is similar to human growth hormone. It contracts with the maternal metabolic system to supply energy to the growing fetus. The presence and function of this hormone is what allows the fetus to be nourished and grow even to the detriment of the mother in the event of malnutrition.

Progesterone is just what is says it is: pro gestation. Without progesterone life will not be sustained. The amount of progesterone hormone produced by the corpus luteum is sufficient for the maintenance of pregnancy, but it must continue to increase in production to sustain the pregnancy throughout. Progesterone now must work double time to increase the blood vessel pathways to the womb, thus increasing blood flow to the womb. This hormone also alerts the endometrium to: begin production of nutrients to support the embryo and to begin the formation of the dicidua where the placenta will attach and the embryo will implant. Once placental attachment has occurred, this hormone will be responsible for its establishment and growth. The placenta will now begin to produce its own hormones, and for much of the rest of pregnancy progesterone will concentrate on keeping the uterus relaxed and not contracting. As amazing as progesterone is, it would not be able to function properly without the presence of estrogen. **Estrogen** is responsible for the overall production and maintenance of all other pregnancy hormones. Estrogen is key in the development of the fetus, especially the vital organs. Also, it is a support hormone which stimulates the growth and helps to maintain the function of the placenta. Estrogen does not cause contractions, but in a way makes them more efficient as the hormone improved the sensitivity of the myometriem improving its function in labor. During pregnancy another key hormone arrives on the scene known as **relaxin**. This hormone will loosen muscles, ligaments and joints, and in earlier pregnancy prepare the body for the growth of the baby. Later, this hormone will equip the pelvic region to soften, open and ready for the passage of baby during the birth process.

The Hormones of Labor

Progesterone

Estrogen

Oxytocin

Prostaglandins

Adrinalin

Endorphins

Progesterone and **estrogen** are found to be quite active during this process as well, but closer to labor progesterone will begin to taper off while estrogen will increase. Remember, progesterone helps to maintain the pregnancy by preventing uterine contractions and estrogen support the efficiency of the contractions. It is important then for progesterone and estrogen to trade places as labor nears so contractions are no longer inhibited, but rather now can be encouraged. Where in the early part of pregnancy estrogen was a supporting hormone to progesterone, it will now play a similar role with oxytocin and later with prolactin. **Oxytocin** will be the hormone of note leading to and during labor. Oxytocin does not actually trigger contractions, but rather as progesterone decreases and estrogen increases the uterus becomes more sensitive and responsive to oxytocin and labor is allowed to naturally progress. Oxytocin will progressively increase and aided by estrogen initiate contractions that will eventually bring baby. Oxytocin is considered by many the love hormone and will aid in the bonding between mother and her baby during the postpartum period. During early labor oxytocin is not only responsible for the uterine contractions, but also for the stretching of the cervix. The stimulation and the ripening of the cervix is usually one of the first indications that progesterone is moving out and oxytocin is now at work. Here is where we are introduced to **prosteglandins** also an integral part in preparation for labor and of the ripening and release process of the cervix. **Endorphins** are natural pain relief hormones and have been equated to the strength of narcotics in their effectiveness to help reduce pain. These endorphins, or hormones, are released as labor is allowed to progress naturally. They will not be released if synthetic hormones are given to augment labor. **Adrenalin** provide much needed energy which will get the mother into labor and also help her be alert immediately following all the hard work of labor. However, this hormone is not welcome in large quantities during the labor process. Adrenalin is also known as the fight or flight hormone and should the mother become afraid or have to "fight" for the birth she has planned oxytocin is inhibited and labor is interrupted. Oxytocin does not stop once the baby has been born, but this hormone remains vital until the placenta is birthed and the uterus contracted. A baby on a breast will stimulate an increased production of oxytocin and oxytocin will encourage contractions and restrict the blood flow to the uterus, which until now has been encouraged and in fact desired.

The Hormones of Lactation

Finally, the hormones of lactation. As we have seen previously, progesterone and estrogen play significant roles, but their levels, which have supported and maintained pregnancy will take a drastic dive, still present, but now **prolactin** is present and the hormone of note. Like progesterone the word itself breaks it down and makes it easy for us to understand just what prolactin does: it is pro lactation. Actually, the pregnant mother is prepared physically to breastfeed around four months, but due to high levels of both estrogen and progesterone lactation is inhibited until the baby is born. Now, after the birth of her baby, the love and bonding promoter hormone oxytocin having peaked and mixed with endorphins which create the perfect cocktail of hormones for milk let down. Prolactin levels will then peak in the immediate postpartum time probably around 20 minutes after birth. Prolactin levels will taper off by about 50% in the week following birth, but will continue to be produced when the nipple is stimulated.

Cultivating Hormones

We can cultivate certain hormones whether they are stress hormones or the opposite ideal. The way we feel and the atmosphere in which we give birth can promote or inhibit the hormones we need for the best outcomes.

Promote oxytocin in the following ways:

Ensure comfort

Stay calm

Trust your care providers

Know your environment

Avoid intrusion

Stimulate nipples (or clitoris)

Promote

Inhibit oxytocin in the following ways:

Fear

Harsh lights

Talking

Others talking

Releasing adrenalin

Promote endorphins in the following ways:

Stay calm

Trust your body

Trust the birth process

Choose comfort

Remain undisturbed

Low lighting

 **Remember!** 2-3 days postpartum can see a drop in hormonal levels and the onset of the baby blues.

Nutrition

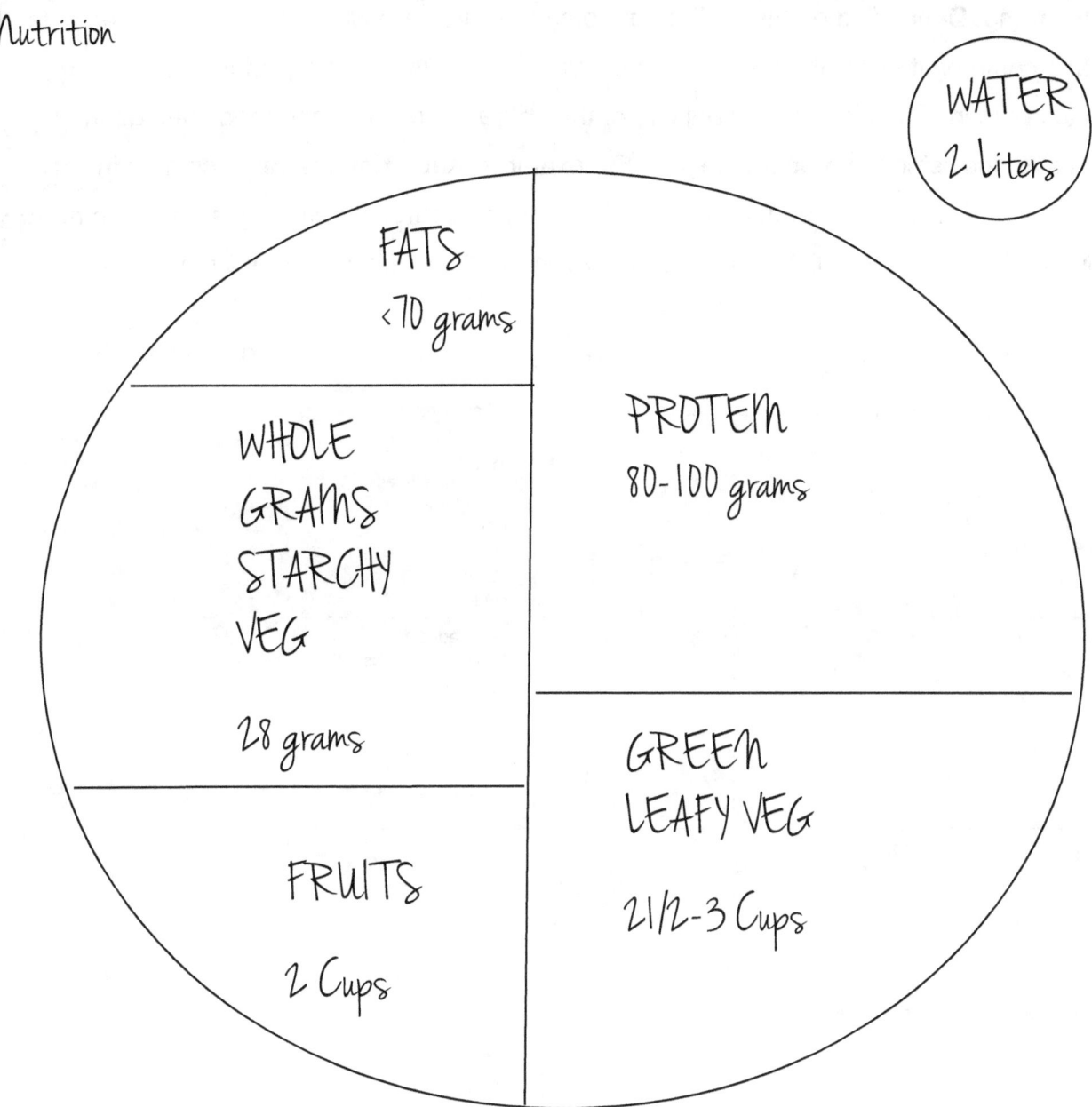

Almost every disease women face in pregnancy can be prevented with good nutrition. Our bodies need both micro and macro nutrients to be well, all of which are found in whole foods. The foundation of a healthy pregnancy is the food we eat. It is imperative we know how to counsel young mothers, or mothers to be, on the best sources of whole foods to ensure proper vitamins and minerals during the development of their baby and into the lactation period.

Research additional whole food sources for each of the following categories and write them in the the space provided. It is very important to know a variety of food sources for each of the following. Depending on the mother, or community you are working in, you may have to replace common items with others due to culture, availability or taste preference. If you plan to serve in a specific country outside of the US research common food sources they may have and list in the proper category. For example: Kang Kong is very common in Vietnam and can be listed under the green leafy or iron source category, plantain, common throughout Africa, can be listed under starchy veg and is a high source of vitamin A.

Nutrient	Pregnant	Breastfeeding	Source	Additional Sources
Protein	80-100grams	80grams	Black Beans Salmon Chickpeas Cottage Cheese Red Meat Chicken Eggs Lentils Quinoa	
Fats	<70 Grams	<70grams	Avocados Sunflower seeds Olive Oil Coconut Oils Nuts Flax Seeds	
Green Leafy Veg	2 ½ -3 Cups		Swiss Chard Kale Spinach Bok Choy	
Whole Grains & Starchy Veg	28 Grams		Oats Brown Rice Millet Plaintain Pumpkin Corn	
Fruits	2 Cups		Apricots Blueberries Cherries Grapes Watermelon	

Nutrient	Pregnant	Breastfeeding	Source	Additional Sources
Calcium	1000mg	1000mg	Milk Yogurt Cheese Spinach Broccoli Kale Figs	
Magnesium	220mcg	320mg	Green Leafy Veg Nuts and Seeds Avocado Banana Fish Dark Chocolate	
Iron	49mg	16mg	Liver Oysters Chickpeas Pumpkin Seeds Beans Lentils Spinach	
Zinc	11 mg	12mg	Oysters Red meat Poultry Beans Nuts	
Potassium			Beans Dark leafy greens Potatoes Squash Yogurt Fish Avocados Mushrooms Bananas	
Vitamin A	770mcg	770mcg	Dark Leafy Greens Sweet Potato Carrots Pumpkin Tomatoes Squash Cantaloupe Red Peppers	
Vitamin D	15mcg	15mcg	Sunshine Poultry Eggs Cheese Beans Fish	
Vitamin E	15mcg	15mcg	Seeds Nuts Oils Swiss Chard Beet Greens	
Vitamin K	90mcg	90mcg	Alfalfa Kale Nettles	

©2016 GoMidwife

Nutrient	Pregnant	Breastfeeding	Source	Additional Sources
Vitamin B 6 and 12	2.6mcg	2.8mg	Pork Chicke Turkey Fish Brown Rice Oatmeal Eggs Milk Cheese Meat	
Folate	600mcg	500mcg	Dark Green Veg Broccoli Chickpeas Beans Lentils	
Vitamin C	85mg	120mg	Berries Citrus Fruits Papaya Tomato Kiwi Broccoli	

Notes _____

Where Does the Weight Go?

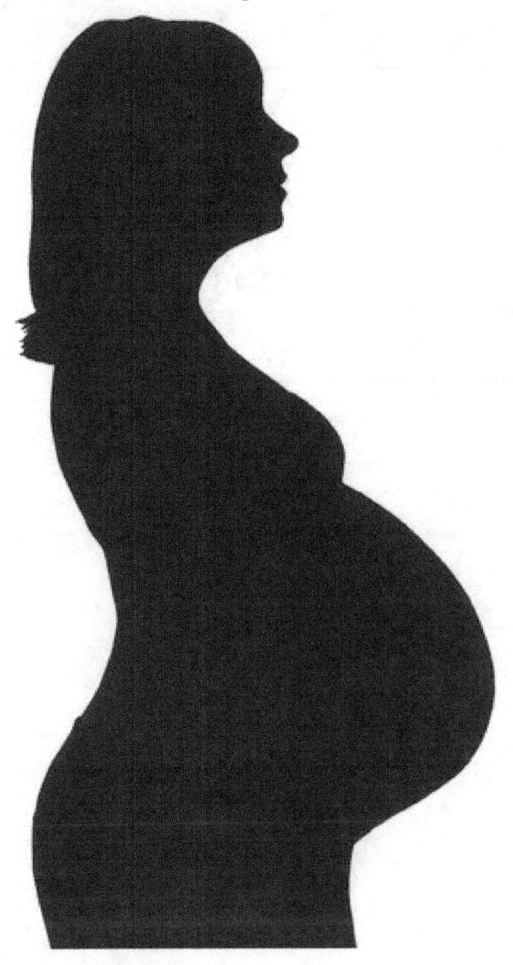

Baby: 7.5 pounds
Placenta: 1.5 pounds
Amniotic fluid: 2-3 pounds
Breast tissue: 2-3 pounds
Blood and Fluid: 4-8 pounds
Stored fat for delivery and breastfeeding: 5-9 pounds
Uterus: 2 pounds
Total: 25-35 pounds

How Does the Time Break Down?

Pregnancy is calculated as 40 weeks, but truthfully, most women will gestate, on average 41 weeks and 3 days, which means, some will be pregnant less than 41 weeks and some more than 41 weeks. It is important we give them this time and not grow impatient or afraid nor in anyway suggest they do so themselves.

Early Term: Between 37 weeks 0 days and 38 weeks 6 days
Full Term: Between 39 weeks 0 days and 40 weeks 6 days
Late Term: Between 41 weeks 0 days and 41 weeks 6 days
Post Term: Between 42 weeks 0 days and beyond

©2016 GoMidwife

Trimester	Month	Weeks
First	One	1-4
	Two	5-8
	Three	9-12
Second	Four	13-16
	Five	17-20
	Six	21-24
	Seven	25-28
Third	Eight	29-32
	Nine	33-36
	Ten	37-42

Notes_____
_____
_____
_____
_____
_____
_____
_____
_____

God has no hands but ours.
Susan Wilson, CPM

Normal Adult Vital Signs

Vital signs are the response of the body's most basic functions. The four most commonly monitored vital signs are:

Blood Pressure

Respiration Rate

Pulse

Temperature

Blood Pressure

Normal Ranges from: 90/60 to 140/90

Average Ranges from: 110/70 to 120/80

Hypertension >140 over >90 is cause for concern.

Hypotension <90 over <60 is cause for concern.

Always take the baseline into consideration.
Posture effects blood pressure. Readings will be highest when sitting and lowest when lying on the left side.

Remember!
Emotions and posture can also effect blood pressure, specifically the systolic number.

Respiratory Rate

12-16 breaths per minutes when resting.

Respirations may increase slightly during labor or at higher elevations.

Rapid breathing can be a sign of: dehydration, infection, fever, shock, anemia or nervousness.

Difficulty breathing in can be a sign of: infection of the heart or lungs or heart failure.

Difficulty breathing out can be a sign of: asthma

Pulse

60-100 beats per minute

The average pulse for a pregnant mother is 80 beats per minute

An irregular pulse can be a sign of heart disease.

A weak pulse can be a sign of poor circulation cause by dehydration, anemia or shock.

Temperature

97.8° F to 99.0° F
36.5° C to 37.2° C

The temperature can be measured:

Orally: by mouth

Axillary: under the arm

When taking an axillary temperature add 1°.

Notes

Routine Prenatal Assessment

Visit	Action	Labs	Counseling and Education
Initial Visit 6-10 weeks	Health History Risk Assessment Height Weight Head-to-Toe Assessment Blood Pressure FHT (10 weeks with doppler)	Prenatal Panel: CBC HIV Rubella GC/Chlamydia Syphilis Hepatitis H/H ABO/Rh/Ab Urine Culture PAP (can be deferred)	Domestic Violence Nutritional Diary Warning Signs Teratogens Herbs in Pregnancy Vitamins and Supplements
12 Weeks	Blood Pressure Fundal Height Weight Pulse Urine Dip FHT Fetal Motion	MSAFP/Triple/Quad Screen Diagnostic Ultrasound	Diet and Exercise Normal Discomforts Emotions in Pregnancy Fetal Development Informed Consent Home Birth Informed Consent MSAFP
16 Weeks	Blood Pressure Fundal Height Weight Pulse Urine Dip FHT Fetal Motion		Sex in Pregnancy
20 Weeks	Blood Pressure Fundal Height Weight Pulse Urine Dip FHT- with fetoscope Fetal Motion		Preparing for HomeBirth/Birth
24 Weeks	Blood Pressure Fundal Height Weight Pulse Urine Dip FHT Fetal Motion Position	GDM/GTT	Nutritional Diary Reassessment Informed Consent GDM
28 Weeks	Blood Pressure Fundal Height Weight Pulse Urine Dip FHT Fetal Motion Position Edema	Hct/Hgb Rhogam	Optimal Fetal Positioning Birth Supplies Water Birth Informed Consent Rhogam
32 Weeks	Blood Pressure Fundal Height Weight Pulse Urine Dip FHT Edema Fetal Motion Position Edema		Fetal Kick Count Fear in Pregnancy Worksheet

©2016 GoMidwife

Visit	Action	Labs	Counseling and Education
36 Weeks **Home Visit**	Blood Pressure Fundal Height Weight Pulse Urine Dip FHT Fetal Motion Position AAT Edema Assess Breast/Nipples	GBS	Signs of Labor Breastfeeding Considering Transport Informed Consent GBS
38 Weeks	Blood Pressure Fundal Height Weight Pulse Urine Dip FHT Fetal Motion Position AAT Edema		Postpartum Care Review Birth Plan
39 Weeks	Blood Pressure Fundal Height Weight Pulse Urine Dip FHT Fetal Motion Position AAT Edema		Newborn Care Informed Consent : NBS Vitamin K Eriythromicin
40 Weeks	Blood Pressure Fundal Height Weight Pulse Urine Dip FHT Fetal Motion Position AAT Edema		Review Dates Discuss Alternative Augmentation
41 Weeks	Blood Pressure Fundal Height Weight Pulse Urine Dip FHT Fetal Motion Position AAT Edema	BPP	Consult Consider Alternative Augmentation

Charting in Antepartum

Health History Form

Name	First	Middle	Last	Maiden		Date	Phone (home) (work)
Race	Religion	Yrs. Educ.	Marital Status	Occupation/Type of Business		Date of Birth	State of Birth
Address: Street			City	Zip		Inside City Limits Yes No	How long at this address?
Father of Baby:	First	Middle	Last	Race	Yrs. Educ.	Date of Birth	State of Birth
Address (if different from above- if the same write "Same")						Phone (work) (home)	Occupation/Type of business
Partner/Husband (if different from Father)		Another person to contact in emergency Name:				Phone: Relationship:	
		E-Mail:					
Social Security #:		Father's Social Security #:		Social Security Number requested for baby? Circle YES or NO		How did you hear about our services?	

Please complete this form in preparation for your initial visit. Your responses will be kept confidential. This information assists us in optimizing your care. If you need more spaces, please use the area provided on the back.

Yes No   Have you or the father of the baby (FOB) ever had a baby with birth genetic anomolies?
Yes No   Are you and the FOB related by blood? (e.g. cousins)
Yes No   Are you or the FOB from any of these ethnic/racial groups? (circle)
                Jewish   Black/African   Asian   Mediterranean
Yes No   Have you or the FOB ever had hepatitis or jaundice?
Yes No   Do you think you are at an increased risk for AIDS/HIV or STD's?
Yes No   Have you ever experienced dramatic fluctuations in your weight?
Yes No   Have you ever had anorexia, bulimia or other eating problems?
Yes No   Have you ever been in an abusive relationship or been abused ?
Yes No   Have you ever had severe emotional problems?
Yes No   Have you ever been on any medication for psychological problems?
Yes No   Has anyone ever told you, or do you think you have ever used alcohol or drugs excessively?
Yes No   Have you ever had a blood transfusion?

Height:_____

Pre-Pregnancy Weight:_____   **Current Weight:**_____

Do you know your blood type? Yes   No If yes, what is your blood type?_____

MEDICAL HISTORY –
Please indicate if you have ever had any of these issues? When?

Severe headaches	Bowel problems/colitis
Eye vision problems	Blood in stool
Ear/hearing problems	Gall bladder problems
Dental problems	Liver problems
Thyroid	Hepatitis
Rheumatic fever	Diabetes

©2016 GoMidwife

Blood clotting problems	Hypoglycemia
Anemia	Bladder infection
Hemorrhage	Kidney infection
High Blood Pressure	Urinary surgery
Varicose veins	Urethral dilation
Hemorrhoids	Aching joints
Tuberculosis	Pelvic/back injuries
Asthma	Seizures
Skin disorders	Cancer
Stomach problems	Hospitalizations
Ulcers	**Surgeries**
Chicken Pox	Fifth's Disease
Trauma	Fractures
Depression	Psychiatric Drugs
Toxic Exposure	Occupational Hazards
Blood transfusion	**Immunizations**

Current Pregnancy

Last Menstrual Period (1st day) _____ Normal? Yes No

Conception Date _____

Did you take a pregnancy test? Yes No Date_____

Planned pregnancy? Yes No

Feelings about this pregnancy
_____

Father's feelings
_____

Most recent birth control used
_____

Contraception used in past; what, when, any problems?
_____

Do you plan more children? Yes No Would you like information in Natural Family Planning? Yes No

Please indicate if you've had any of the following problems prior to pregnancy, during this pregnancy or in previous pregnancies:

Nausea	Yeast
Heartburn	Vaginal Infections (other)
Fatigue	Group B Strep
Urinary Tract Infections	Cervical Surgery
Herpes	Syphilis

Gonorrhea	Chlamydia
Pelvis Infections	Abortion
Fibroid	Uterine Surgery
Breast Surgery	Ovarian Cysts
HIV/AIDS	STD (other)
Headache	Vomiting
Abdominal Pain	Vaginal Bleeding
Fever	Depression
Constipation	Vaginal discharge
Other	Other

Gynecological History

Age of first period cycle_____ Cycle length_____ Are your cycles regular? Yes   No

Do you have pain during menstruation? Yes   No

When was your last clinical breast exam?_____

When was your last PAP smear?_____ Have you ever had an abnormal PAP? Yes   No

Have you been pregnant before? Yes   No   How many pregnancies have you had?

Have you ever had a miscarriage?   Yes   No   If yes, when? _____

Have you had a C-Section? Yes   No   If yes, when_____

Have you had a VBAC? Yes   No   If yes, when _____

How many babies did your mother have?_____

Vaginal or C-Section?_____

# Pregnancy and Date of Pregnancy	Length of Gestation	Length of Labor	Male or Female Single or Twin	Comments on Pregnancy

Sexual History

Do you have pain with intercourse?   Yes   No

©2016 GoMidwife

How many sexual partners have you had?_____

Are you in a monogamous relationship? Yes   No

Has your partner had previous sexual relationships or current ones other than you?

What is your activity level?
_____

Tell me about your
nutrition:_____
_____
_____
_____

Do you have any allergies to medications? Yes   No   If yes, please list_____

Do you have any allergies to latex? Yes No

Are you taking any medications? Yes No
If yes, please list the medication, the dosage and the reason for each medication:
_____

Are you currently taking any herbs ? Yes   No   If yes, please list:
_____

Please indicate if you have used, experienced, or been exposed to any of the following during this pregnancy:

Tobacco	Herbs
Alcohol	Fumes or Sprays
Caffeine	X-ray
Ultrasound	Cocaine
Marijuana	Measles
Street Drugs	Vaccinations
Cats	Other

Is there anything about yourself you think I should know or that you would like me to know so that I might serve you and your family better?

_____

Midwife   Date   Reviewed/Date

©2016 GoMidwife

Prenatal Record

Client ID#_____
DOB_____ Age_____
Partner's Name_____

Height_____
Pre-Pregnancy Weight_____
Weight at Birth_____

Name_____
EDD_____
G_____ P_____

Allergies:_____
Review Dates

Comments:

LNMP	
Fetal Movement	
FHT Fetoscope	
FHT Doppler	
Ultrasound	

Weeks Gestation	12	16	20	24	28	32	34	36	38	40	41	42
Date												
Visit												
Weeks Gestation												
Fundal Height												
Position												
Weight												
Urine Analysis												
Blood Pressure												
Pulse												
Fetal Movement												
Fetal Heart Rate												
Edema												
Vaginal Exam												
Labs												
Meds												
Diet/Exercise												
Complaints/ Concerns												
Midwife's Initials												

©2016 GoMidwife

Notes

Head to Toe Assessment

Name

Date

Date of Birth

EDD

Blood Pressure

Pulse

Respirations

Temperature

Urinalysis

General Appearance

Head

Skin

Eyes

Ears

Tongue

Gums

Clavicles

Neck

Heart

Breasts

Lungs

Liver

Abdomen

Fundal Height

Fetal Heart Rate

Reflexes/Clonus

Edema

S

O

A

P

Prenatal Face Sheet

Name	
Date of Birth	
Address	
Phone	
E-mail	
EDB	
LNMP	
Social Security #	
Partner's Name	

Date	Lab	Result	Date	Lab	Result
	H/H (initial)			Serology	
	H/H (28 wk)			Glucose (28 wk)	
	Blood Type			Rubella	
	Antibody			Pap	
	Antibody			GBBS	
	HBsAG			GC	
	HIV			Chlamydia	
	MSAFP			Urine Culture	

U/S Date & Place	Gestational Weeks & Findings

Action	Date	Initials
Informed Consent, Practice Disclosure, Fee Policy discussed and signed.		
Exercise and Diet Worksheet given on:_____ Evaluated on:_____		
MSAFP discussed 15-18 weeks: Yes:_____ Result:_____ Declined:_____		
Rh Antibody Screen (if negative blood type): 24weeks:_____ 36 weeks:_____ Declined:_____		
Rhogam given: 28 weeks:_____ 48 hrs postpartum:_____ Declined:_____		
Glucose Tolerance Test given 28-32 weeks:		
GBS swab 36 weeks:		
Home Visit 36 weeks:		
Childbirth Education Classes discussed.		
Plans to breastfeed: Yes_____ No_____ (Check for inverted nipples.)		
Birth Kit		
Newborn Informed Consent given and signed: Vitamin K: Yes_____ Declined_____ Erythromycin: Yes_____ Declined_____ NBS: Yes:_____ Declined_____		
Emergency Transport Plan discussed.		
Postpartum Care Plan discussed.		
Birth Certificate Worksheet given and discussed.		

Antepartum Notes

ANTEPARTUM NOTES

Page _____

Client ID # _____
Name _____
Partner's Name _____
Allergies _____
Blood Type _____
EDD _____

Date	Weeks	Notes
Initials		Signature

©2016 GoMidwife

Understanding Common Labs

There are certain labs which are considered routine in the United States. Below are a list of the most common labs to be drawn in pregnancy.

Prenatal Panel: 10-12 weeks This will likely be the first lab you draw on a new mom. It is best to draw this lab on the second visit, and allow the first, or initial, visit to be one of relationship building. This test is a blood screen and will look at the mothers overall health. Routinely included in this panel are the following:

Complete Blood Count (CBC)-this test will look at the white blood cells, the red blood cells and the platelets. White blood cells are present in greater numbers when a bacteria or infection is also present in the body. It also looks at the red blood cells and these are the oxygen carriers. They carry oxygen from the lungs throughout the rest of the body and carbon dioxide from the body back to the lungs to be exhaled. When the red blood count is low then anemia is present. Lower levels of red blood cells means neither the mother nor the baby will get the oxygen they need to thrive. Platelets are the smallest part type of blood cell and is responsible for clotting. Platelet counts are important to know for the postpartum period.

Hematocrit and Hemoglobin (HCT/Hgb)- hematocrit measure the red blood cell volume, or how much space the red blood cell is taking up in the entirety of the blood itself. The average values for hematocrit are **34.9 to 44.5 percent for women** which means if your hematocrit is 38 then 38% of your total blood volume are red blood cells. The hemoglobin indicates the ability of the blood to carry oxygen. Hemoglobin is also what gives the blood its color and is the filler for the red blood cell. Normal values are usually **12.0 to 15.5 grams per deciliter**. Hemoglobin values are generally 1/3 the value of the hematocrit so if you have a hemoglobin value of 11 then the hematocrit will likely be 33.

Rubella- This test determines whether or not a woman has been exposed to the German Measles

Syphilis- One of the most prevalent STD's in the past is quite rare today unless the woman has a high risk lifestyle. If she does have this disease it can be devastating for both her and the baby.

ABO- This test the woman's blood type and is determined by the type of molecule which sits on the red blood cell. A person can have A (a antigens), B (b antigens), AB (both a and b antigens), or O (neither a nor b antigens) and this signifies their blood type.

RH- If a mother has a specific antigen the she is Rh+ of the antigen is absent then she is Rh-.

Antibodies- Antibodies are only important if the mother is a negative blood type and does not have antigens present in her bloodstream.

HIV- The CDC recommend all pregnant women be tested for HIV, but it can be declined.

Triple Screen/Quad Screen: 15 weeks This is a genetic test provided at approximately 15 weeks. There is a good deal of controversy surrounding this test as it has a reputation for many false positives. This test is not a diagnostic test, rather a screening. Should the screening come back positive, further tests would need to be run. This is a blood draw.

Glucose Tolerance Test (GTT) 25-28 weeks- This test is routine to determine whether or not a woman has an inability to properly process sugars during pregnancy. There is some question as to whether or not this test needs to be routine or only when a women is at risk. This test is performed after a woman takes a download of sugar and has a finger prick test.

Group B Strep (GBS) 35-38 weeks- This is a vaginal and rectal swab to determine whether or not the naturally occurring bacteria is present at the time of the test. This bacteria is transient and often a woman will test positive prior to birth who do not have if at birth and vice versa.

Urinalysis

A mother's nutritional levels, hydration as well as organ function can all be determined at a glance of a urine dipstick. A person's urine is very telling to their overall health. Understanding how to read a urine dip stick accurately is a vital role to quality prenatal care and can alert you to potential problems before they are chronic.

Prior to dipping the urine you will want to make an overall assessment. Note the following:

Color
Odor
Turbidity

Color- Urine color can vary greatly and for various factors. Normal urine color ranges from clear, or no color, to dark yellow. There are a few colors that can be noted which would be cause for further investigation. Most commonly variations range from brown, red, orange to dark yellow and often mean dehydration or poor or impaired kidney function. There are also benign causes of discoloration, which include food, vitamins and medications. Be sure to rule out the most likely cause before jumping to the worst case scenario.

Odor- when assessing urine for odor you are assessing for signs of infection. Urine does have a smell and at times this normal smell can be strong, but should not be considered pungent or foul. Fruity or sweet smelling urine can also be an indication for concern and often indicates diabetic ketoacidosis.

Turbidity- is the noted cloudiness of the urine sample. It can be cause by a variety of reasons including mucous or excessive vaginal discharge, which can be common in pregnancy, as well it can indicate the presence of STD's, dehydration, and more often it is an indication of an infection in the urinary tract which includes the kidneys, bladder and urethra.

When assessing urine with a urinalysis strip be sure to fully submerge the dip stick in the urine then turn the stick on its side on a paper towel to allow excess fluid to drain off and the prevention of cross contamination from tab to tab. The main assessment is for dehydration, signs of preeclampsia, urinary tract infections including bladder and kidney, malnutrition and signs of diabetes. You will note the following:

Leukocytes
Blood
Bilirubin
Protein
Keytones
Spefic Gravida
Glucose
Urobiligen
Nitrates

It is vital to know what each of these indicate, and more specifically, what they indicate in the pregnant woman. Urinalysis is a science unto itself and quite complex. Here is what you need to know:

Leukocytes-The presence of leukocytes, or white blood cells, in the urine is very common in pregnancy and trace amounts are often present due to increased vaginal secretions which often become mixed in with the urine sample. When higher levels are present, or when leukocytes are present along with nitrates or blood, then infection should be considered. The most common to consider in the pregnant woman are: bladder, kidney, yeast (not associated with the presence of blood or nitrates) and asymptomatic bacteriuria.

Blood- Blood in the urine is a cause for concern. It likely will indicate some type of infection, such as a UTI, or could imply the presence of kidney stones. However, blood in a term pregnancy could simply be a sign of cervical change.

Bilirubin- Increased levels of bilirubin in the urine could indicate liver insufficiency or damage. Rarely seen in the urine bilirubin is the by-product of the breakdown of red blood cells and does not generally present in the urine.

➤**Protein**- Trace amount of protein in the urine is absolutely normal and not a cause for concern. However, spilling protein is one of the key indicators of impaired kidney function and one of the first signs we might be given that a women is preeclamptic. If you note more than a trace of protein in the urine this can be a cause for further diagnosis. Protein in the urine is the main reason to routinely sample urine in pregnancy.

Keytones- In the non pregnant woman, keytones present in the urine could be a sign of diabetes, but in the pregnant women, trace or small amounts of keytones more likely points to a need for more calories. Keytones are present in the urine when fats are broken down, instead of carbohydrates, for energy. Higher levels of keytones found in the urine can indicate malnutrition or dehydration and need to be corrected at once.

Spefic Gravida- This measures the concentration, or the amount of particulates in the urine. Most commonly, a high concentration indicates dehydration and the need for the consumption of more water.

Glucose-Sugar is rarely detected in the urine. Should your urinalysis indicate glucose in the urine, then further diagnosis is necessary to exclude the possibility of gestational diabetes.

Urobiligen- Trace amounts are possible, but the presence of higher levels of urobiligen in urine is very rare and mostly likely indicates a need for new dip sticks. However, if you are sure your sticks are accurate then liver and/or kidney impairment would be the most likely cause for high levels of urobiligen present in urine and further diagnosis would be necessary. Keep in mind, you can get a false positive on urobiligen when high levels of nitrates are present.

Nitrates- The presence of nitrates in the urine implies the presence of bacteria, especially when present with blood and/ or leukocytes, indicates an infection, mostly likely a UTI infection.

pH- Normal urine is generally measured at a 5.5 - 6.5, which is slightly acidic to neutral. When measuring the pH of urine it is based on the scale of 1 -14 with 1 being acidic and 14 alkaline. < 6 implies acidity and >6 alkaline. Most likely a slight move one way or the other is caused by recent food consumption but urine that is severely acidic can be an indicator of diabetes and severe alkaline can indicate a bacterial infection. Alkaline urine, especially when associated with the presence of keytones can indicate starvation. Changes in the urine pH is rare, pregnancy included.

Blood Sugar Values

Time of Test	Sugar Level
Before eating	Less than 95
One hour after eating	Less than 140
Two hours after eating	Less than 120

Notes_____

_____
_____
_____
_____
_____
_____
_____
_____

Such love has no fear, because perfect love expels all fear. If we are afraid, it is for fear of punishment, and this shows that we have not fully experienced his perfect love.
1 John 4:18

Fear in Pregnancy.

Pregnancy is a time when emotions come to the surface. This can be such a healing time and as a midwife, this is likely the greatest task and opportunity with which we are presented. Fear can impede labor from coming on at all, it can stall labor once it does come, and it can cause poor recovery in postpartum, but we have hope...we do not need to fear. One of the greatest gifts we have as a midwife and one of the sharpest tools we carry is that of councilor and intercessor. Interceding for each and every woman will change their lives. There are many fears they will encounter:

the fear of miscarriage
the fear of morning sickness never going away
the fear of something they eat or do will harm the baby
the fear something will be wrong with the baby
the fear of complications in pregnancy or birth
the fear of pain
the fear they will not be able to breastfeed
the fear they will not be a good mother

All of these fears are real and we should validate each woman and how she feels, but we cannot leave her in that place. We must walk with her to a new vantage point where the light of knowledge and understanding can cast out the unknown darkness. Acknowledge her fears and talk through each one. Encourage her to read material that is beneficial and not fear mongering. Help her to own her own experience and recognize the strength within her. Do your best to explore all fears prior to birth so she can go into the birthing process unafraid.

Fear is the opposite spirit of love, because we are told perfect love cast out all fear. When a woman labors in truth and in the way she was designed to give birth, it is a clear picture of the gospel...one who lays down their life, in full surrender, so another might live. The world would have us look to the fears, the inadequacies, the pain, but if we can help each woman to look past that to the victory, the resurrection and the joy we can change outcomes in pregnancy. We must labor with the woman prior to the time she physically goes into labor. If we will labor in intercession we will find so much of the work is done prior to the day of birth, and on that day not only is a baby born, but she too is born anew in perfect love.

No Fear!

The LORD is my light and my salvation-- whom shall I fear? The LORD is the stronghold of my life-- of whom shall I be afraid? Psalm 27:1

"Giving birth should be your greatest achievement not your greatest fear." Jane Weideman

Be strong and courageous. Do not be afraid or terrified because of them, for the LORD your God goes with you; he will never leave you nor forsake you." Deuteronomy 31:6

"There is a secret in our culture, and it's not that birth is painful. It's that women are strong."
Laura Stavoe Harm

For I am the LORD, your God, who takes hold of your right hand and says to you,
Do not fear; I will help you.
Isaiah 41:13

"You block your dream when you allow your fear to grow bigger than your faith."
– Mary Manin Morrissey

"Nothing in life is to be feared. It is only to be understood. "~Marie Curie

"Fear can be overcome only by Faith." ~Grantly Dick-Read, M.D.

"It is not only that we want to bring about an easy labor, without risking injury to the mother or the child; we must go further. We must understand that childbirth is fundamentally a spiritual, as well as a physical, achievement. The birth of a child is the ultimate perfection of human love."
~Dr. Grantly Dick-Read, 1953

"When you have come to the edge of all light that you know and are about to drop off into the darkness of the unknown, FAITH is knowing one of two things will happen: There will be something solid to stand on or you will be taught to fly" ~Patrick Overton

I am leaving you with a gift—peace of mind and heart. And the peace I give is a gift the world cannot give. So don't be troubled or afraid. John 14:27

When I am afraid, I put my trust in you. Psalm 56:3

Fetal Positions²

Fetal position is determined by the four quadrants of the mother's pelvis and, most commonly, the occiput (the back of the head) on of the fetus, but sometimes will be determined by the fetus's sacrum if the baby is in the breech position.

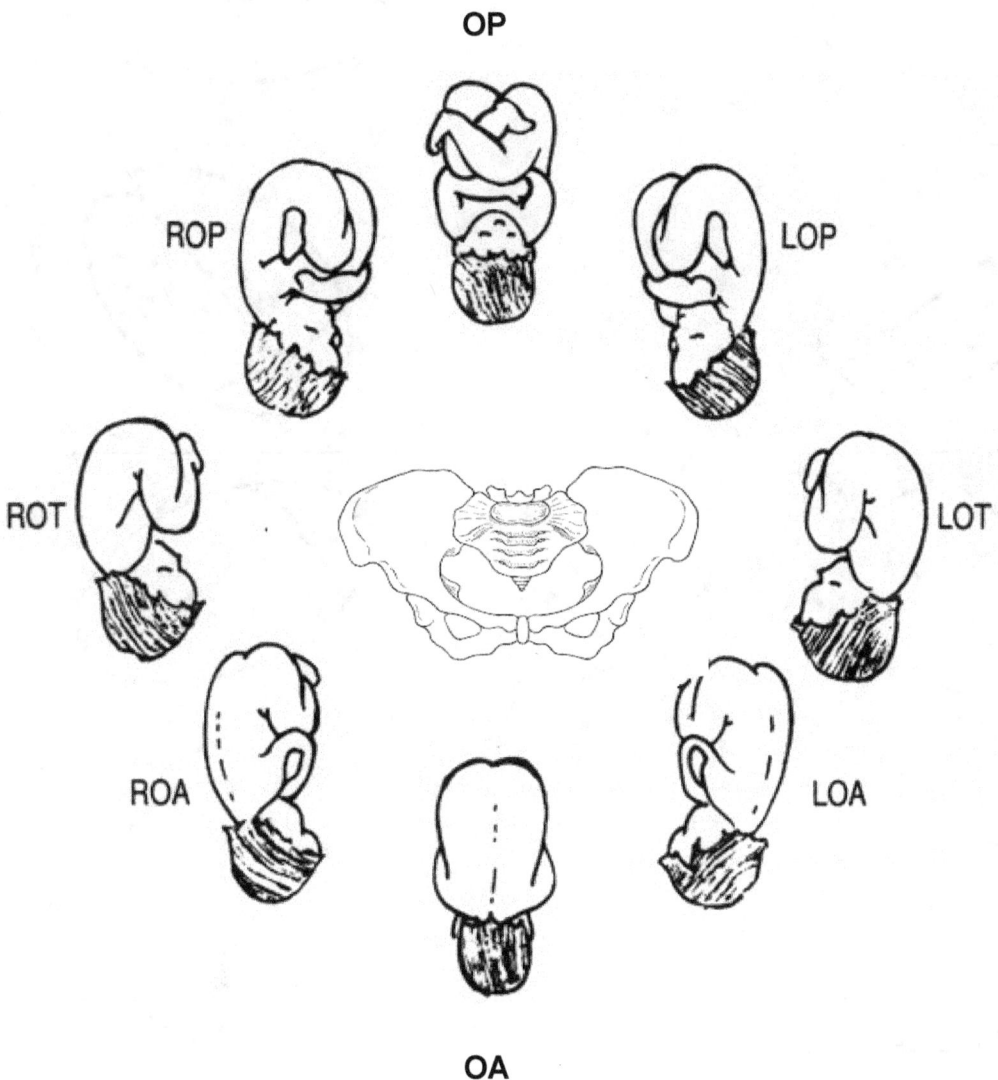

Fetal Positions per Vaginal Exam

Imagine you are performing a vaginal exam. Using the anterior and posterior fontanelles as landmarks, what position is each baby in?

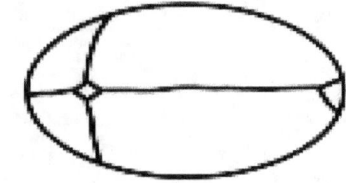

1. _____

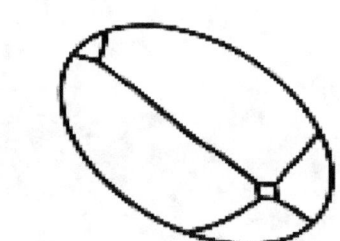

2. _____

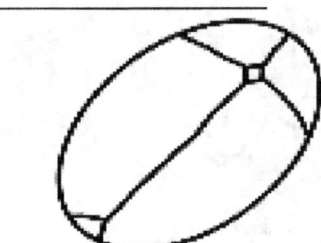

3. _____

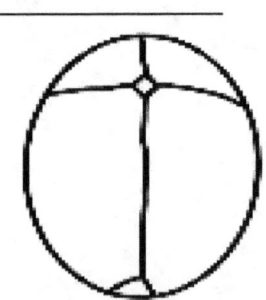

4. _____

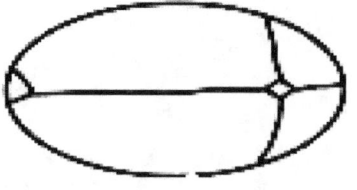

5. _____

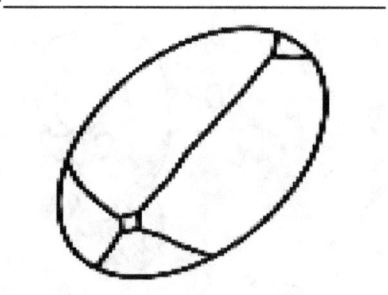

6. _____

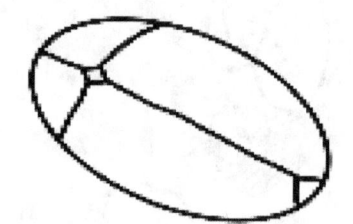

7. _____

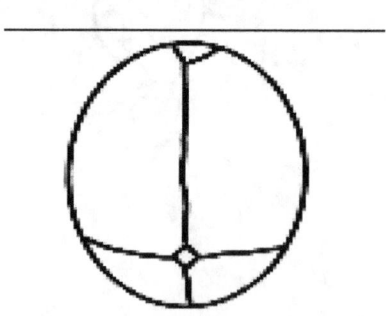

8. _____

©2016 GoMidwife

Notes

Leopold's Maneuvers[3]

Maneuver One: Fingers toward the xiphoid process, palpate the fundus. Identify the fetal part in the upper pole of the uterus. Most likely you will find a bottom or a head.

Maneuver Two: Walk fingers down the sides of the uterus to determine the fetal back.

Maneuver Three: Grasp the presenting part between the thumb and first two fingers to confirm maneuver one.

Maneuver Four: Face the lower pole of the uterus. Determine attitude of baby. If occiput is felt opposite the back, baby is flexed. If occiput is felt on the same side as fetal back, baby is extended.

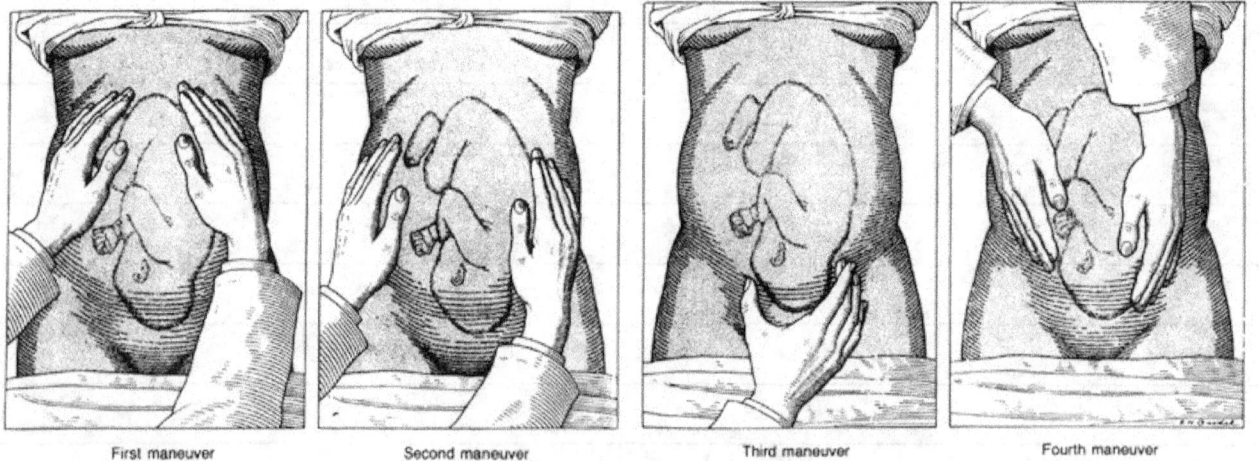

First maneuver   Second maneuver   Third maneuver   Fourth maneuver

Finding Fetal Heart Tones

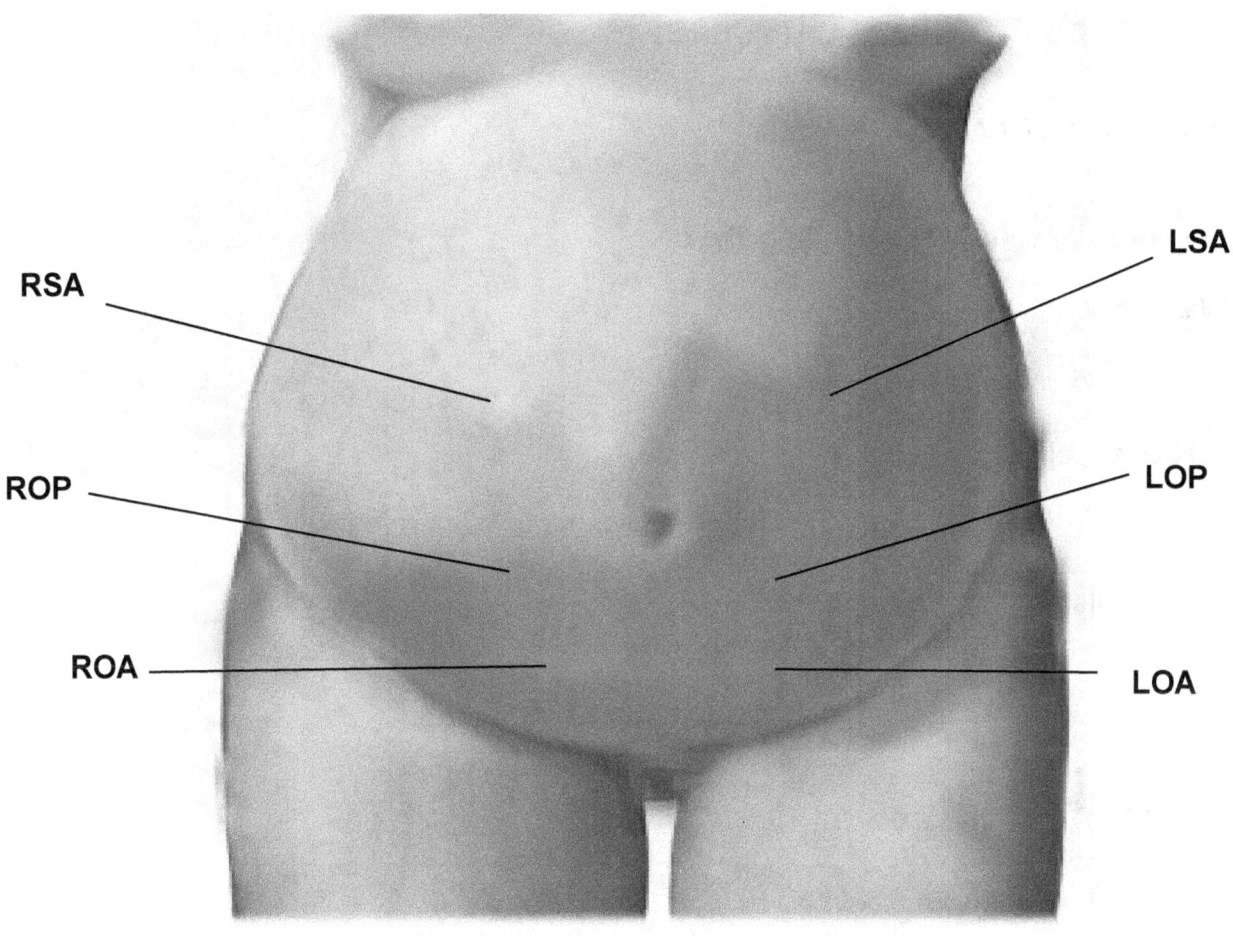

When the Birth Assistant Arrives

Enter the home quietly and respectfully. Remove shoes as appropriate and greet the family quietly. Maintain a servants heart at all times. You are here to serve the mother and the family.

Birth and Postpartum Checklist

Upon Arrival and Prior to Birth

- ☐ Help midwife bring in supplies from the car
- ☐ Wash hands
- ☐ Set up and check birth supplies:
    - NRP supplies
    - Oxygen tank
    - Medication
    - Blankets
    - Towels
    - Trash
    - Laundry
    - Birth kit and supplies
    - Bowl with wash cloths
- ☐ Begin steeping postpartum herbs
- ☐ Prepare peri bottles
- ☐ Review birth plan
- ☐ Review NRP
- ☐ Take and record maternal vital signs accordingly

- ☐ FHT taken and recorded appropriately
- ☐ Keep the mother hydrated and encourage her to void every 1-2 hours
- ☐ Initiate and maintain documentation and flow sheets
- ☐ Observe birthing environment
    Note candles needing blown out if oxygen is necessary
    Drafts of wind or fans
    Where is washer dryer
    Phone
- ☐ Review important information including address in event of emergency transport
- ☐ Take notes

Water Birth

- ☐ Prepare water pots on the stove and maintain simmer
- ☐ Watch mother in the tub and scoop debris accordingly
- ☐ Ensure adequate warm and dry towels for mother
- ☐ Continue to offer fluids
- ☐ Once baby is born maintain body temperature by re-wetting the towels in the warm water and reapplying to exposed skin.
- ☐ Maintain observation on baby's nose and mouth so it does not dip under water
- ☐ Assist in moving mom from tub for placenta delivery

During and Following Birth

- ☐ Note time of pushing, crowning, birth of head and birth of body
- ☐ Call 30 seconds
- ☐ Check cord for pulse or use stereoscope call out the rate
- ☐ Call one minute

©2016 GoMidwife

- ☐ Maintain warm and dry towels for mother and baby
- ☐ Remove wet towels from baby
- ☐ Call five minutes
- ☐ Assign APGAR for comparison with the midwife later
- ☐ Note the time of placenta
- ☐ Monitor stability of mother and baby
- ☐ Offer fluids and nourishment
- ☐ Assist midwife with assessment and suturing
- ☐ Prepare for transition from the tub or birthing area
- ☐ Maintain hygiene removing chux pads and cleaning sheets, plumping pillows
- ☐ Assist in hygiene and shower or sponge bath
- ☐ Assist mother to empty bladder
- ☐ Assist with newborn exam
- ☐ Maintain routine observation and vitals on mother and baby
- ☐ Assist mother with latch
- ☐ Make an appointment for 24 hour phone call and 48 follow up visit

Postpartum Duties

- ☐ Use disinfectant wipes to wipe down all surfaces including stethoscopes, Doppler, etc.
- ☐ Clean dirty instruments, dry and place in Ziploc for transport
- ☐ Put supplies away
- ☐ Drain tub
- ☐ Discard trash
- ☐ Start laundry

- ☐ Check supplies for restocking
- ☐ Place supplies by back door for transport to car
- ☐ Maintain vital assessments on mother and baby appropriately
- ☐ Assess placenta

Assist with Suturing

- ☐ Obtain Lidocaine, and suture kit
- ☐ Assist with sterile field
- ☐ Maintain light source
- ☐ Maintain fundal checks and routine vitals as midwife sutures

24 Hour Phone Call

- ☐ Baby has had at least three successful breastfeeding sessions since departure
- ☐ Baby has urinated
- ☐ Baby has had bowel movement
- ☐ Bleeding is less than 1 pad er hour

48 Hour Home Visit

- ☐ Newborn Screen drawn and mailed
- ☐ Assessment of mom and baby
- ☐ Assessment of breastfeeding
- ☐ Weigh baby and determine % loss since birth
- ☐ Notify midwife of any variances from normal

There is no way out of the experience except through it, because it is not really your experience at all but the baby's. Your body is the child's instrument of birth.
Penelope Leach

Water Birth

Water birth is an ideal way for women to birth, especially first time mothers. Utilizing a birthing tub has many benefits. Included among them are:

Allows mobility
Reduces the length of labor
Reduces blood pressure
Allows for privacy and prevents inhibitions
Reduces perineal trauma
Provides relaxation
Increases pain relief and is known as the home-birth midwife's "aquadural"
Eases baby into your world

One of the risks of water birth is that baby will be too cold once born. In order to prevent this from occurring you should maintain the water temperature at 97°-100°.

Maternal dehydration is also a concern and can be avoided by insuring the laboring woman drinks plenty of water during labor.

Getting in to the tub too early can relax the mother to such a degree that labor slows down and works against you as the provider and her as the laboring mother. Be sure not to get into the tub until you are sure active labor has begun. During a long active labor stage be sure to have the laboring woman get out of the tub every hour and re-enter only after she has been out of the tub at least 30 minutes. This takes place to ensure optimal efficiency of the tub benefits.

Breathe. Focus. Perform. Thrive

Dr. Melanie Harker and Brigid Malloy, CNM

Thinking On Your Feet

A midwife absolutely must be able to think on her feet. No birth is ever the same and you will always need to adapt what you know to each particular woman and labor. You will need to prepare for every situation and be ready to respond appropriately. Here are some tips to use as a beginning midwife to better learn how to think on your feet:

- Don't just react- think first, take a moment to breathe, and think again.
- Slow down- you will perform better if you slow down and work consistently.
- Observe- always be thinking ahead. There is a lot of down time in birth, never just sit there. Always assess the situation and ask what needs to be done next to ensure progression and well-being.
- Always prepare for common complications- prior to birth think through common complications. If you work with a birth team take time to work through scenarios on a consistent basis. This will ensure you each know how to respond appropriately.
- Focus on what needs to be done- relax. Its hard to imagine you can relax during a birth, especially when an emergency occurs, but you can and it will help not only you, but the mother, the family and your birth team.
- Go with your gut- you have been trained for this. Breath, focus and perform. Have confidence in yourself and and trust your instincts. Hesitate only long enough to breath and focus, then....perform.

Notes

One is constantly having to balance the high expectations of modern health care with the need to respect the human soul. This is especially so with birth.
Benig Mauger

Signs of Labor

Lightening or Engagement is when the baby drops further down into the pelvic outlet. Often the mother will be able to breath more easily, but will have increased pressure vaginally. Also, due to the uterus resting on the bladder the need to urinate will more frequently occur. This process can happen several days to several hours prior to the onset of labor. It is usually more evident in first time mothers.

Increased discharge is when the body begins to prepare for labor through lubricating the vagina. Increased discharge can sometimes be confused with the mucous plug. An increase often occurs several weeks to days prior to labor beginning.

Bloody Show is when mucous covering the cervical opening is expelled. The mucous covers the cervix to prevent foreign substances from entering and causing infection. As the cervix ripens and effaces, opens and dilates the mucous plug will come out. Sometimes there will be just a bit of mucous and sometimes the entire plug will come out at once. This can occur several weeks prior to labor, and can be might be tinged pink or yellow. like someone blew their nose.

Contractions is when the body begins to prepare for labor in the weeks leading up to the actual event. Braxton Hicks contractions are part of the process. They are irregular contractions with no real pattern and do not usually last with a change in position. Contractions that signal impending labor are regular, progressively stronger and closer together, are longer and follow a pretty clear pattern. They will radiate from front to back or back to front.

Ruptured Membranes are when the amniotic sac is ruptured. Usually contractions come prior to this even,but labor can begin with the waters breaking. Once waters break, labor generally occurs within 24 hours. They can break all at once or a small trickle. Fluid should be clear or slight tinged yellow.

Normal Labor Patterns

First Time Mother

Latent Labor 8-12 hours
Active Labor-6 hours not usually more than 12
Pushing- 1 hour no more than 3

Prolonged Labor- 20 hours +

Second Time Mother

Latent Labor- 5 hours
Active Labor-3 hours not usually more than 6
Pushing- 20 minutes no more than 1 hour

Prolonged Labor- 12 hours +

Prolonged labor us generally due to: cervix, uterus or pelvis. To be able to adjust, it is important to know the cause of the prolongation.

Water Birth often decrease the first stage of labor from 468 minutes to 380 minutes on average. Decreasing the first stage of labor approximately 88 minutes. Water birth also shows an average decrease of second stage of labor to 34 minutes.[4]

Failure to progress is failure to wait
Abigail Callahan

Stages of Labor

First Stage - this stage is from the onset of labor until the cervix has effaced and dilation is is complete at 10 cm. It can last 8-20 hours, but typically we expect 12-15 hours for first time mothers laboring naturally. This stage is broken into three phases:

Early Labor
although dilation does occur, effacement, or the thinning of the cervix, is the primary action during this time. The longest phase of labor, early labor, generally lasts from **8-12 hours**. Contractions will be anywhere from **5-30 minutes** apart and last **45-60 seconds**. This is a time to rest. If labor begins in the day time, use this time to complete last minute baby preparations and then rest, sleep and relax. Take a warm bath, go see a movie, hydrate and conserve energy. If it occurs at night, sleep is the best option. The hardest work is yet to come.

Active Labor
this phase will typically take 6 hours. It begins with the cervix approximately **4 cm.** Dilated and will be complete when the cervix reaches **8-9 cm**. Contractions become stronger, longer and more intense during this phase. You will begin to see a shift to about **3-4 minutes apart lasting about 60 seconds**. This is often the hardest part of labor.

Transition
this phase can be anywhere from 30 minutes – 2 hours with contractions between 2-3 minutes apart lasting about 1 minute. This period of time is often accompanied by the mother making verbally irrational statements. Nausea and vomiting can also accompany this stage as well a marked spacing between contractions to allow the mom to rest for impending second stage.

Second Stage- begins with pushing and is complete when the baby is born. Pushing can last from **30 minutes to 3 hours**. Crowing is when the widest part of the baby's head is visible.

Third Stage- delivery of the placenta can take from **15 minutes to 1 hour** to separate and deliver. It is noted the placenta has separated when the cord lengthens are a gush of blood is noted.

Notes_____
_____
_____
_____
_____
_____
_____
_____
_____
_____
_____
_____
_____
_____
_____
_____
_____
_____
_____

Cervical Dilation

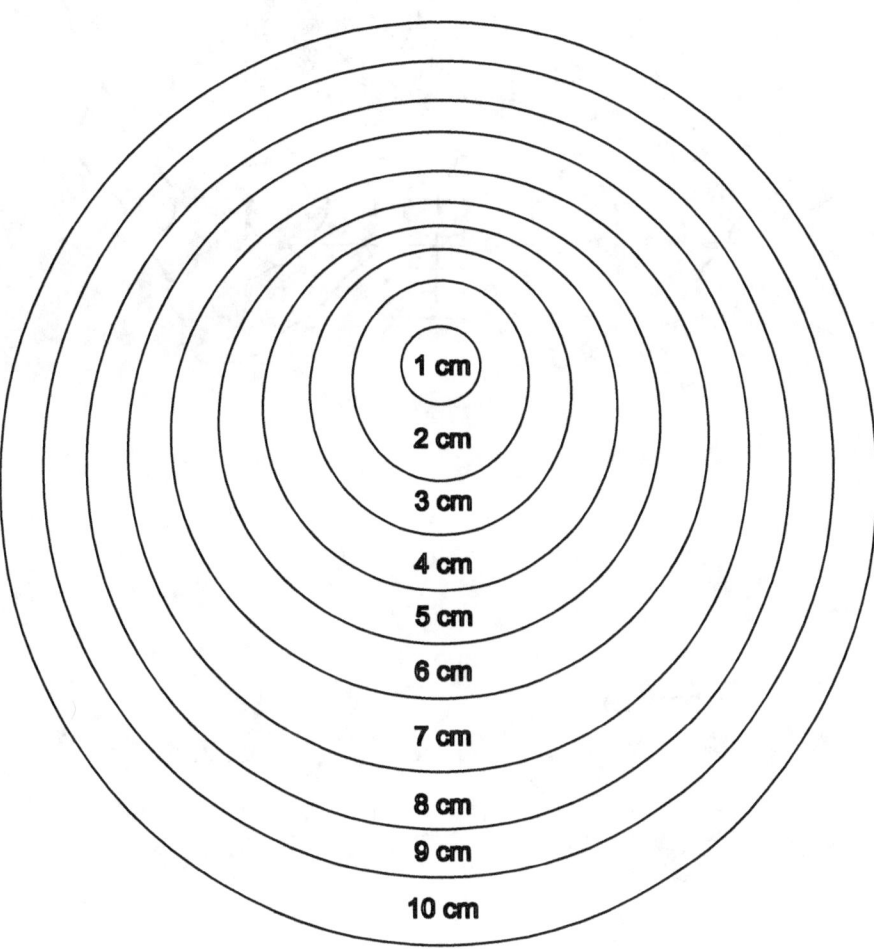

Station[5]

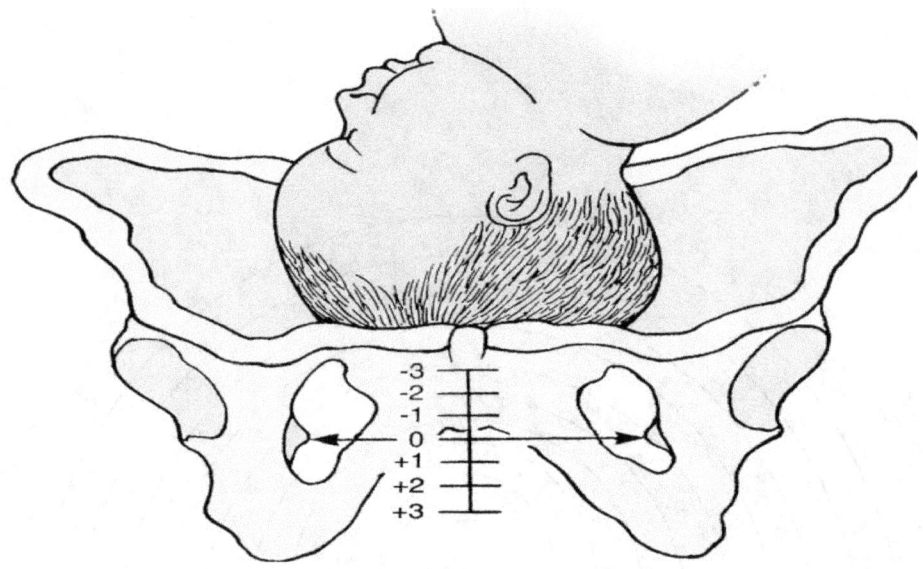

Effacement[6]

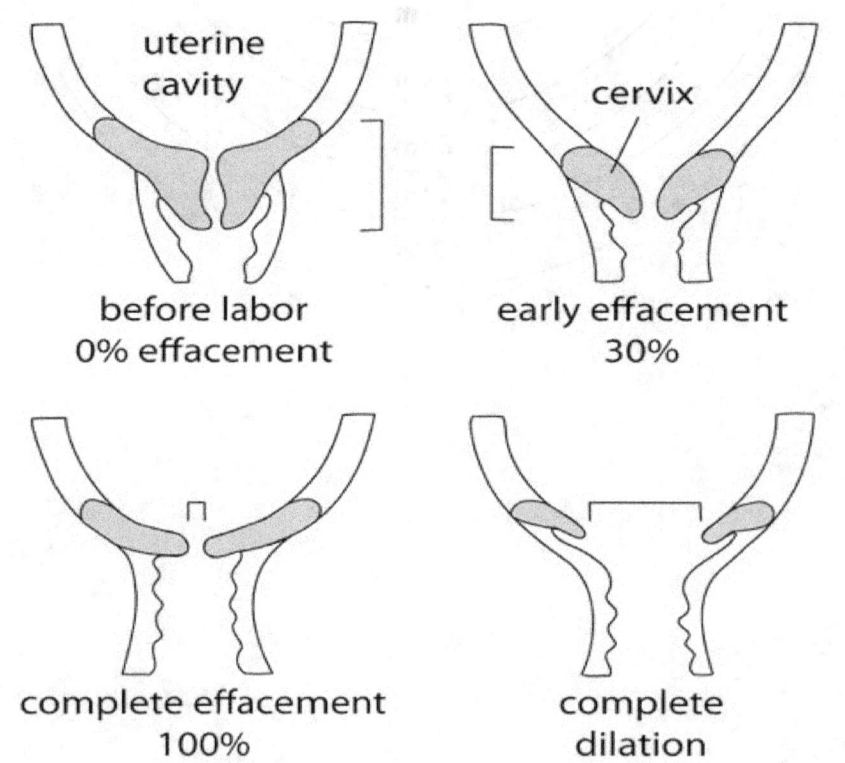

Notes

Charting in Labor

Labor Flow Sheet

Name:														Date:	
Time															
Temp															
Pulse															
Resp															
BP															
FHR Baseline															
Regular / **I**rregular															
Decreases: Ø = none **A**brupt / **G**radual															
Increases Present															
Uterine resting tone: **S**oft / **F**irm															
Maternal Position/Activity															
Initials															

Comments:

Midwife Signature _____ Initials_____

Birth Assistant Signature _____ Initials_____

©2016 GoMidwife

Client ID#_____ Age_____
Name_____
Partner's Name_____

Allergies_____
Blood Type_____ G____ P____
EDD_____

Labor Flow

Page_____

Significant Prenatal
History:_____
Onset of Labor:_____
_____
_____
_____

Progress Notes Date:_____

Time	Vital Signs	Ctx. Freq. Duration	I & O	FHT	Notes	Initials

Signature_____ Initials_____
Signature_____ Initials_____

©2016 GoMidwife

Times

Times in labor are very important. Although every labor plays out on its own time and at its own pace, there are normal limits to consider. Time seems to stand still in labor and an hour can some times feel like 15 minutes. Be sure to always mark significant times in labor and throughout its progression. Here are a few you will want to note:

Onset of contractions
Time midwife arrives
Onset of active labor
SROM
Onset of transition
When the mother is "pushy"
Onset of pushing
Visualization of head
Crowing
Birth of head
Birth of body
Birth of placenta
Time cord is cut
Time of newborn's first latch and protein feed
Time midwife departs

Notes_____
_____
_____
_____
_____
_____
_____
_____

APGAR

Activity/Muscle Tone

0- Absent
1- Arms and legs extended
2- Active Movement

Pulse/Heart Rate

0- Absent
1- Below 100
2- Above 100

Grimace/Irritability

0- No Response
1- Facial Grimace
2- Sneeze, Cough, Pull Away

Appearance/Color

0- Blue/Gray, Pale All Over
1- Pink Body/Blue Extremities
2- Pink Over Entire Body

Respirations/Breathing

0- Absent
1- Slow, Irregular
2- Good, Crying

Vital Signs and Assessment

During Labor:

Mom	Baby
Temperature, BP, Respirations and Pulse every 4 hrs.	FHT every hour in early labor
More frequent if abnormal	Complete AAT graph upon arrival
Temperature every 2 hours once SROM occurs	FHT every 30 minutes during active labor
Resting uterine tone with every FHT	FHT every other contraction during stage two
	FHT continuously if decels are noted and until they spontaneously resolve or transport is initiated.

Following Birth:

Mom	Baby
Assist skin-to-skin	Straight to mom's chest
Complete a set of vitals at 15 minutes postpartum	APGAR at 1 and 5 minutes
Once placenta delivers complete vitals, including fundal check, every 15 minutes for the first hour	Follow NRP protocol for out of hospital birth
Vitals, including fundal check every 30 minutes for the second hour	Keep baby dry, warm, skin-to-skin
Vitals every hour until departure	Complete vitals at 15 minutes of life (following initial assessment)
Mom should void within 30 minutes postpartum	Complete vitals every 15 minutes during the first hour of life
	Complete vitals every 30 minutes during the second hour of life
	Complete vitals every hour until departure
	Respirations should be assessed for one full minute on newborns

Over assessment is always better than under assessment.

Newborn Note and Exam

Mother's Name _____ Client ID# _____

Father's Name _____ Date: _____ Time: _____

Newborn's Name: First _____ Middle _____ Last _____

Midwife: _____ Assistant: _____

☐ Male

☐ Female

Date of Birth: _____

Time of Birth: _____

Location of Birth: ☐ Home ☐ Birth Center ☐ Other

Cord Blood Yes No
Blood Type: _____

Resuscitation

☐ None
☐ Stimulation
☐ Blow By
☐ PPV Bag and Mask _____ minutes
☐ CPR

Suction

☐ None
☐ Bulb
☐ DeeLee

Breastfeeding: ☐ Yes ☐ No
Time of First Protein Feed: _____ Date: _____

Birth Summary: _____
_____

1 Min	5 Min	APGAR Score
0 1 2	0 1 2	Heart rate Absent < 100 >100
0 1 2	0 1 2	Respiratory Rate Absent Weak Cry/Gasp Strong Cry/Breathing
0 1 2	0 1 2	Reflex Absent Grimace Cough/Sneeze
0 1 2	0 1 2	Muscle Tone Limp Some Flexion Well Flexed
0 1 2	0 1 2	Color Pale Body Pink/Extremities Blue All Pink

Vitamin K

☐ Declined
☐ Administered
   ☐ Injection
   ☐ Oral

Eye Ointment

☐ Declined
☐ Administered

©2016 GoMidwife

Time of Exam:_____
Examined By:_____
Estimated Gestational Age:_____

Assessment:

Weight:_____
Length_____in. _____cm.
Head Circumference: _____ in. _____cm.
Chest:_____in. _____cm

General Appearance:_____
Color:_____
Temperature:_____

HEENT:_____
Heart: ❏ No Murmur_____
Lungs/Respiration:_____
Abdomen:_____

Cord: 3 Vessels   ❏Yes ❏ No

Hips: ❏Full ROM ❏No Clicks_____
Genitals: _____
Anus: ❏ Terminal Mec ❏ Patent_____

Back/Spine: ❏No Dimple ❏No Tufts_____

Skin: ❏   BirthbMark(s)_____
Tone:_____

Reflexes: ❏Suck ❏Root ❏Moro ❏Plantar ❏Palmar ❏Babinksi ❏Galant

Notes:_____
_____
_____
_____
_____

Midwife's Signature_____ Date_____ Time_____

©2016 GoMidwife

Reflexes

Testing a newborn's reflexes soon after birth can give us insight into the health of baby, along with the development of specific muscles and neurological connections. These reflexes are often tested within the first couple of hours of baby's life, more specifically during the newborn exam. The following are the reflexes tested during a newborn exam:

<u>Moro/startle</u>- tested by positioning baby in a seated position while supporting the head. The person testing will allow baby to fall backward slightly, catching him almost instantly. If the reflex is active/present, baby will lift palms upward with thumps out when startled and will bring arms back into torso when caught.

<u>Rooting</u> - tested by stroking baby's cheek. If reflex is present/active, baby will turn toward the check that was stroked and should being to make suckling motion.

<u>Sucking</u>- tested when mother introduces nipple into baby's mouth. If reflex is present/active, baby will instinctively suck anything that comes in contact with the roof of his mouth.

<u>Babinski</u>- tested by stroking the bottom of baby's foot from the toe to the heel. If the reflex is present/active, baby's toes will fan out and the big toe will move upward.

<u>Grasp</u>- tested by placing one finger into the palm of baby's hand. If the reflex is active/present, baby will grasp the finger and maintain a firm grip.

<u>Step</u>- tested by holding baby upright as if walking. If the reflex is present/active, baby will place one foot in front of the other, as if walking, when his foot comes into contact with a flat surface.

Notes

Ballard Gestational Exam[7]

Neuromuscular Maturity

Score	-1	0	1	2	3	4	5
Posture							
Square window (wrist)	>90°	90°	60°	45°	30°	0°	
Arm recoil		180°	140°–180°	110°–140°	90°–110°	<90°	
Popliteal angle	180°	160°	140°	120°	100°	90°	<90°
Scarf sign							
Heel to ear							

Physical Maturity

Skin	Sticky, friable, transparent	Gelatinous, red, translucent	Smooth, pink, visible veins	Superficial peeling and/or rash; few veins	Cracking, pale areas, rare veins	Parchment, deep cracking, no vessels	Leathery, cracked, wrinkled
Lanugo	None	Sparse	Abundant	Thinning	Bald areas	Mostly bald	
Plantar surface	Heel-toe 40-50 mm: –1; <40 mm: –2	>50 mm, no crease	Faint red marks	Anterior transverse crease only	Creases anterior 2/3	Creases over entire sole	
Breast	Imperceptible	Barely perceptible	Flat areola, no bud	Stippled areola, 1–2 mm bud	Raised areola, 3–4 mm bud	Full areola, 5–10 mm bud	
Eye/Ear	Lids fused loosely: –1; tightly: –2	Lids open; pinna flat, stays folded	Slightly curved pinna; soft; slow recoil	Well curved pinna; soft but ready recoil	Formed and firm, instant recoil	Thick cartilage, ear stiff	
Genitals (male)	Scrotum flat, smooth	Scrotum empty, faint rugae	Testes in upper canal, rare rugae	Testes descending, few rugae	Testes down, good rugae	Testes pendulous, deep rugae	
Genitals (female)	Clitoris prominent, labia flat	Clitoris prominent, small labia minora	Clitoris prominent, enlarging minora	Majora and minora equally prominent	Majora large, minora small	Majora cover clitoris and minora	

Maturity Rating

Score	Weeks
-10	20
-5	22
0	24
5	26
10	28
15	30
20	32
25	34
30	36
35	38
40	40
45	42
50	44

Newborn Exam SOAP

Age at Exam

Height

Weight

Head Circumference

Chest Circumference

Temperature

Respirations

Heart Rate

General Appearance

Activity and Cry

Color and Skin

Head: Shape, Fontanels, Sutures

Eyes: Lips, Conjunctiva, Sclerae, RR

Ears: Placement, Cartilage, Hearing

Nose: Shape, Patency, Nares

Mouth: Lips, Palate, tongue, Gums, Mucosa

Neck: ROM, Clavicles

Chest: Breasts

Heart: Rhythm, +/- Murmur, Regular/Irregular

Lungs: Rate, Sounds, Pattern, Bilateral

Abdomen: Contour, Skin, Umbilicus, BS, Hernia

Femoral Pulses

Back: Spine, Dimple, Hair/Tufts, Open/Closed

Extremities: Arms, Hands, Fingers, Legs, Hips, Feet, Toes

Rectum/Anus: Describe, Stool to Date

Genitalia: Labia and Vagina, Penis and Testes

Void

Reflexes: Moro, Root, Suck, Palmar, Grasp, Plantar, Babinski, Stepping

Assessment

Plan

Notes

Pounds to Grams

	0	1	2	3	4	5	6	7	8	9	10
0	0	454	907	1361	1841	2268	2722	3175	3629	4082	4536
1	28	482	936	1389	1843	2296	2750	3203	3657	4111	4564
2	57	510	964	1417	1871	2325	2778	3232	3685	4139	4593
3	85	539	992	1446	1899	2353	2807	3260	3714	4167	4621
4	113	587	1021	1474	1928	2381	2835	3289	3742	4196	4649
5	142	595	1049	1502	1956	2410	2863	3317	3770	4224	4678
6	170	624	1077	1531	1984	2438	2892	3345	3799	4252	4706
7	196	652	1106	1559	2013	2466	2920	3374	3827	4281	4734
8	227	680	1134	1588	2041	2495	2948	3402	3856	4309	4763
9	255	709	1162	1616	2070	2523	2977	3430	3884	4337	4791
10	283	737	1191	1644	2098	2551	3005	3459	3912	4366	4819
11	312	765	1219	1673	2126	2580	3033	3487	3941	4394	4848
12	340	794	1247	1701	2155	2608	3062	3515	3969	4423	4876
13	369	822	1276	1729	2183	2637	3090	3544	3997	4451	4904
14	397	850	1304	1758	2211	2665	3118	3572	4026	4479	4933
15	425	879	1332	1786	2240	2693	3147	3600	4054	4508	4961

Signs and Symptoms of Respiratory Distress

RDS usually occurs at birth or in the hours immediately following. < 3 hours

Tachypnea- Rapid or labored breathing > 60 breaths per minute.

Grunting on exhale

Flaring nostrils

Retractions- when the skin pulls inward between the ribs or under the ribcage

Circumoral Cyanosis- blue skin around the nose and mouth.

Inspecting the Placenta

It is very important to do a thorough examination of the placenta immediately following birth. The placenta can give you a view and further insight into the overall health of the pregnancy. You should always assess for: completeness, size, color, odor, cord insertion sight and any significant anomalies. On average the placenta will weigh approximately 1/6th (divide by 6) to that of the weight of the baby. If you note a much larger placenta then it is possible the mother may have had undiagnosed gestational diabetes, much smaller placentas can be a sign of hypertension.

The older the placenta, the duller the color and you may see increased calcium or salt deposits. You will also want to know whether or not there has been meconium staining of the placenta.

If there is a depression in the placenta or an adherent clot, placental abruptions may have taken place.

The placenta should have the same earthy smell as a period or postpartum bleed and should not smell foul in any way. If you note an odor, infection would be the likely assessment.

Always pull the membranes together to note the presence of one whole and to confirm all membranes are present.

Assess the cord for: insertion sight, appearance, length and diameter. The color of the cord is an off white, or ecru. The length of cord is on average 55-60cm and it is important to also include the portion still connected to the newborn. Some midwives prefer to cut the cord long, and this should be factored into the measurement. A few cords will measure <35cm and some >80cm, but the average is stated above.

Note any variations and send to pathology when possible if unknown anomalies are present.

Examination of the Placenta

- ☐ Explain the procedure to the parents and ask if they want to observe.
- ☐ Wash hands and put on a pair of clean gloves.
- ☐ Lay out the placenta with the fetal surface facing up – note shape, size, color and smell
- ☐ Examine the cord. Note and record: length, point of insertion and presence of any knots
- ☐ Count the vessels in the cut end of the cord. Note two arteries and one vein.
- ☐ Observe the fetal surface for irregularities.
- ☐ Lift the cord and hold placenta up, observe the membranes and inspect for completeness. There should be a single hole present.
- ☐ The placenta is returned to the surface and the membranes are spread out in order to look for extra vessels, lobes, or holes in the surface.
- ☐ Separate the amnion and the chorion to ensure that they are both present.
- ☐ The placenta is turned over to inspect the maternal side.
- ☐ Examine cotyledons to ensure that they are all present, note any areas of infarction or blood clots
- ☐ Place placental in a gallon plastic bag, seal and dispose accordingly
- ☐ Estimate total blood loss at this time
- ☐ Draw cord blood if appropriate.
- ☐ Clean away equipment
- ☐ Remove gloves
- ☐ Wash hands

Notes

Birth Note

Course of Labor- spontaneous, augmented, type and progress

Times of Complete Dilation

Time Beginning to Push

Type of Birth: SVD or Other

Live female or live male

Over intact perineum or lacerations and degree

Presentation and Position

Nuchal cord and management

Suctioning if required

Shoulders and management

Resuscitation management if needed or infant to mothers chest

Time of cord clamping and cutting

Cord blood collected and time

Typing of newborn blood and time

Time of placenta and management

Intact placenta and # of vessels

Presenting: Duncan or Schultz

Abnormalities of the placenta or cord

EBL

Anatomy Inspected: vagina, perineum and cervix

Laceration degree

Well appropriated without bleeding or repaired and type of suture

Infant weight and APGAR

Mother and baby current condition

Postpartum care

Assessment

Plan

Normal Newborn

Temperature

Axillary
97.3°F -99.3°F
36.5°C-37.4°C

Always add 1° when an axillary temperature is taken.

Both low and high temperature readings can be concerning in a newborn.

Respiratory Rate

30-60 breaths per minutes

Newborn breathing is mostly irregular.
Listen for one full minute.

Heart Rate

110-160

Tachycardia >160
Bradycardia <110

Bradycardia is ominous in a newborn and generally indicates hypoxia

Weight

5.5lbs – 10lbs.

2494.76g – 4535.92g

Head Circumference

32-37 cm

12.5-14.5 inches

The head should be approximately 2cm greater than the chest measurement.

Chest Circumference

30-35 cm

12-14 inches

Length

20 inches

50 cm

Blood Sugar

Full-term: 40 and 150 mg/dL.
Premature: 30 and 150 mg/dL.

Fontanelle

Anterior Fontanelle

3-4 cm long x 2-3 cm wide, diamond shaped

Posterior fontanelle

1-2 cm at birth, triangle shaped

Notes

Signs and Symptoms of Infection

Poor feeding
Breathing difficulty
Listlessness
Decreased or elevated temperature
Unusual skin rash or change in skin color
Persistent crying
Unusual irritability
Redness, streaking, odor or heat around umbilical stump

A marked change in a baby's behavior, such as suddenly sleeping all the time or not sleeping much at all can also be an indication that something is not right. These signs are an even greater concern if baby is less than 2 months old.

Normal Newborn Behavior

Sleeping- newborns sleep 16-17 hours a day, but not usually more than 2-3 hours at a time, sometimes 4 hours. The first 24 hours of life will likely find the baby very alert, but following the initial alert period baby will begin to sleep quite a lot. As long as baby is not lethargic when awake and actively nurses well with consistent pees and poops, then sleep should be welcomed for all.

Eating- baby should nurse every 2-3 hours during a 24 hour period which equls to 8-12 feedings per 24 hours.

Peeing- a newborn may urinate only once in his or her first 24 hours. A pattern will likely be established by day six and baby should be urinating eight to 10 times a day. Before the breast milk comes in it is common to see "brick dust" in a diaper. It is pink in color or red and is consecrated uric acid crystals. This should not be seen after day 4 or after the milk comes in.

Pooping- baby will likely poop within the first 24 hours of life. This will be black, sticky tar consistency called meconium. As baby transitions, poop will be green, light brown and then a yellow seedy color.

Once it turns yellow seedy, it should remain this color for breastfed infants. Breastfed babies do not poop as often as bottle fed infants as the breast milk is almost entirely absorbed. White mucous or red blood in stool is a cause for concern.

Common Newborn Skin Conditions

Find a picture of pink pimples. Print and paste it here.	**Pink Pimples-** also known as baby acne are thought to be caused by exposure in the womb to maternal hormones. No treatment is needed, just time. They can last for weeks or even months on baby's skin.
Find a picture of Erythema Toxicum. Print and paste it here.	**Erythema Toxicum-** is another common newborn rash. It looks like red splotches with ill-defined borders that are slightly raise. This rash may also present with small white or yellow dots in the center. Its cause is unknown and it resolves without treatment after a few days or weeks.
Find a picture of Milia. Print and paste it here.	**Milia-** are tiny white bumps or yellow spots found on the cheeks, chin, nose or forehead of the newborn. They are caused by skin-gland secretions. This common rash generally disappears on its own in the first two to three weeks of life.

Find a picture of peeling skin. Print and paste it here.	**Peeling Skin-** can be seen in almost all normal babies, but it is especially noticeable in babies born a little late. The underlying skin is perfectly normal, soft and moist.
Find a picture of salmon patches. Print and paste it here.	**Salmon Patches-** these are also called strawberry patches, or stork bits. They are most commonly found on the forehead around the eyes, or on the back of the neck. These are clusters of blood vessels, probably caused by maternal hormones. They will fade on their own in time. These can take longer to fade than the others. Anywhere from a few weeks to a year of life.
Find a picture of Mongolian spots. Print and paste it here.	**Mongolian Spots-** are very common in any part of the body, but particularly the back and buttocks.. Mongolian spots are more common on dark skinned newborns. They are flat, gray-blue in color and look very similar to a bruise. They can be large or small, dark or faint. They are harmless and often fade away within the first few years of life, but not always.

©2016 GoMidwife

Notes _____

_____
_____
_____
_____
_____
_____
_____
_____
_____
_____
_____
_____
_____
_____
_____

Normal Newborn Pee and Poop Chart

Kidneys

Age of Newborn	Amount	Normal Color Spectrum
First 24 Hours	1 Wet Diaper	In the first day or 2 urine can be light orange or dark yellow. This indicates the presence of bilirubin and nitrates. As the milk comes in
First 48 Hours	2 Wet Diapers	and baby becomes more hydrated urine will transition.
Day 3	3 Wet Diapers	It is possible to see brick dust in a diaper when mother's milk has not yet come int. This will occur around the 3rd- 4th days and is pinkish orange. It can look like blood and is always concerning to parents.
Day 4	3-4 Wet Diapers	Now the milk is in the urine will quickly begin to dilute due to increased hydration.
Day 5	6-8+ Wet Diapers	Pale urine that is rarely visible to the eye.
Day 6	8+	Pale urine is a sign of good hydration.

Bowels

Age of Newborn	Amount	Normal Color Spectrum
First 24 Hours	1 Diaper	Meconium is sticky black stool first present. Meconium is a combination of amniotic fluid, skin cells, lanugo and bile.
First 48 Hours	1-2 Diapers	
Day 3	2-3 Diapers	Stool will begin to transition from meconium. If baby is jaundices, often a good bowel movement will clear it up right away.
Day 4	3-4 Diapers	As breastfed babies transition the stool will be yellow, seedy and creamy or pasty.
Day 5	3-4+ Diapers	
6 Weeks+	Babies may have a BM after every feeding, or may go a day or two between BM's. As long as pee diapers remain the same and poo remains yellow and seedy this is normal.	

This chart is to be used as a guide. It is not exact. **Use as a guide only**.

Lochia Chart

Days Postpartum	Amount	Normal Color Spectrum
First 24 Hours	Less than one full pad per hour.	Lochia rubra is the bright read discharge immediately after childbirth and consists of: mucous, leukocytes, dicidua and blood.
First 48 Hours	Less than one full pad per hour.	Lochia serosa can range in color from pink to brown. It contains less blood and more leukocytes. This consist mainly of leukocytes from the wound site as well as mucous from the cervix.
72 Hours		
Week 1		
Week 2		Lochia alba is the last transition is usually whitish to clear in color and consists mainly of dicidua cells, leukocytes, epithelial cells and cervical mucous.

What is not Normal:

- Heavy bleeding that soaks through a heavy overnight pad within one hour.
- Lochia that doe not transition from rubra to serosa or alba by 8 days after childbirth.
- Smells strong and foul
- Is accompanied by fever and/or chills
- Is accompanied by pain in the power abdomen
- Returns to rubra after having transitioned to alba

 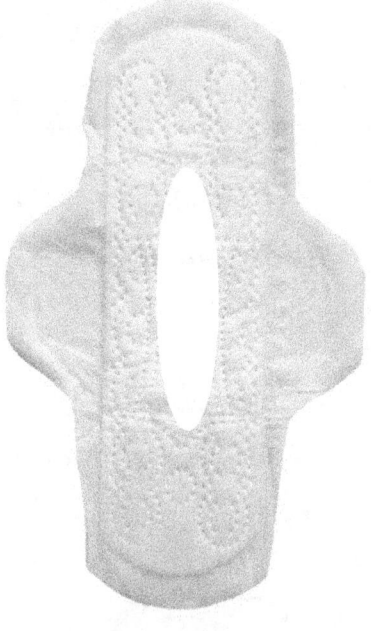 

Small/Scant
Less than 4 in.
Fills pad every 3 + hours

Moderate
Less than 6 in.
Fills pad every 3 hrs.

Heavy
Saturated in
1 hr. or >

Notes_____
_____
_____
_____
_____
_____
_____
_____
_____
_____
_____
_____
_____

Postpartum Visit

Maternal	Newborn
Blood Pressure	Temperature
Temperature	Respirations
Pulse	Heart Rate
Respirations	
Hydration/Food	Birth Weight
Uterus- Involution	Today's Weight
Pain	
Breasts – enlarging with increasing milk supply, soften after feedings	# Voids
	#BM
Nipples – inverted, intact, without pain, elongated after feeding but not pinched or abraded	HEENT – symmetrical, clear, red reflex, moist, fontanels flat
Uterus – firm, measuring at _____	Heart – regular, no murmurs
Lochia – rubra, small to moderate amount, without clots	Lungs – respirations easy, clear
	Abdomen – soft, no masses
Perineum – without tenderness	Umbilicus – healed
Appetite - eating well, plenty of fluids	Hips – symmetrical, no click
Voiding - without difficulty	Genitalia – appropriate, testes descended
BM – denies constipation or fear	Extremities – full ROM, pink, pulses strong
Legs – negative Homan's sign	Lumbar – straight, no dimple
Emotions – happy, bonding, occasionally tearful	Reflexes – intact, strong
	Appearance – pink, tone appropriate
Provide Appropriate Education	Assessment of optimal latch:

Note Variations from Normal	Wide open mouth?
	    Flanged lips?
	    Chin pressed into breast?
	    Low lip covering larger portion of areola?
	    Rhythmic suck/swallow pattern?
	    Content after feedings?
	    Non-painful tugging of the breast?
	    Visually seen colostrum?
	    Drowsiness when nursing?

©2016 GoMidwife

Notes

A newborn baby has only three demands. They are warmth in the arms of its mother, food from her breasts, and security in the knowledge of her presence.
Breastfeeding satisfies all three.
Grantly Dick-Read

Breastfeeding & the Optimal Latch

Understanding hunger cues can significantly change breastfeeding for a new mom. Crying is not the first hunger cue, and if baby can be comforted and nourished prior to being upset success will likely be greater.

Hunger Cues:

-Rolling eyes

-Smacking

-Rooting

-Sucking on fists

-Fidgeting

Position is Key:

-Nose to nipple

-Tummy to tummy and skin to skin

-Bring baby to breast not breast to baby

-Keep baby's body in align: ears, shoulders and hips

-Aim nipple to the roof of the baby's mouth

-Baby should have head tilted slightly back and chin should not be against the chest

Latch:

-Ensure wide open mouth prior to latching

-Tickle the upper lip to elicit a wide mouth

-Place as much aureola in the mouth a possible

-Lips should be flanged like a fish, not tucked

Signs of a Good Latch:

-Ears wiggle

-You can hear swallowing sounds

-baby's tongue can be seen when you pull down the bottom lip

-It does not hurt longer than the initial latch on

-Nipple is rounded after feeding, not like a tube of lipstick

Kellymom states, regardless of what the position or latch looks like, there are only two real measures of a good latch: is it effective and is it comfortable. Encourage every mother to relax and trust her body and intuition during breastfeeding just like they did in birth[2].

Notes_____

_____
_____
_____
_____
_____
_____
_____
_____
_____

---

[2] http://kellymom.com/ages/newborn/bf-basics/latch-resources/

Packing Your Birth Bag

The contents of your birth bag are as much an area of personal preference as they are a major tool in your tool belt. Below you will find a list of several of the tools most commonly used during birth.

Suggestions for Your Beginning Birth Bag:

-Waterproof doppler

-Stethoscope

-BP cuff

-Watch

-Measure Tape

-Gestational Wheel

-Fetoscope

-Newborn Scale

-Urine dip sticks

-Bulb Syringe

-Gel

-Clean gloves

-Sterile gloves

-PPV

-Small flashlight

-Pens

-Massage carrier oil: Coconut or olive oil are great, just remember not to add essential oils until you've checked with mom.

-Rebozo

-Honey sticks: can be very helpful for mom to boost energy

-Rice sock: Use heated in microwave while applying pressure

-Journal: can be used to note significant moments in the birth to be given to mom with a congratulatory card

-Personal hygiene and care options: toothbrush/paste, feminine care products, mints/gum, water bottle, good calorie source & change of clothes

You may find you use all of these tools, you may find you use none. As you begin attending birth, make notes about what you would like to have had at your disposal and about what you feel you would like during your own birth!

Notes

Notes

Notes

Notes

Notes

There are two great days in a person's life - the day we are born and the day we discover why.

William Barclay

Footnotes

Kadari, Tamar. "Shiphrah: Midrash and Aggadah", *Jewish Women's Archive*, last accessed September 12, 2015 at http://jwa.org/encyclopedia/article/shiphrah-midrash-and-aggadah.

Kellymom.com, Last accessed August 3, 2015 at http://kellymom.com/ages/newborn/bf-basics/latch-resources/.

NIV Study Bible, Romans 8:36.

End Notes

1. Fetal Development, pp. 66

Drawings used by permission © *1995 - 2014 . The Nemours Foundation/KidsHealth®.*

Conceptual text ideas inspired by parents.com and babycenter.com

2. Fetal Positions, p. 111

Drawing attributed to the *Oxam-Foote Human Labor and Birth* (5th ed) (p. 59) Norwalk, Conn., Appleton-Century-Crofts, 1986.

3. Leopold's Maneuvers, p.114

Drawing from Pritchard JA, MacDonald PC: *William's Obstetrics*, 16th ed. New York, Appleton-Century-Crofts, 1980.

4. Water Birth, p. 124

Thoeni, Zech, Moroder, and Ploner. "Review of 1600 water births. Does water birth increase the risk of neonatal infection?", *Journal of Maternal-Fetal and Neonatal Medicine*, 2005 May; 17(5):357-61.

5. Station, p. 128

Drawing of stage two from Prenatal Yoga Center, ©2014 Jelsoft Enterprises Ltd.

6. Effacement, p. 128

Drawings from pregnancycompanionapp.com, Embracer Innovations ©2014.

7. Ballard Gestational Exam, p. 140

Drawing from *The Journal of Pediatrics*, 119(3), J.L. Ballard, J.C. Khoury, K. Wedif, C. Jarg, B.L. Walsman, and R. Lipp, "New Ballard Score Expanded to Include Extremely Premature Infants." © 1991 by Mosby, Inc.

For Further Reading

The following additional sources might be of interest to you in your own research.

Hormones:

http://www.yourhormones.info/topical_issues/hormones_of_pregnancy_and_labour.aspx

http://www.fastbleep.com/biology-notes/16/79/492

http://www.urmc.rochester.edu/encyclopedia/content.aspx?ContentTypeID=85&ContentID=P01220

http://pregnancy.familyeducation.com/first-trimester/2-weeks-2-days.html

http://www.yourhormones.info/topical_issues/hormones_of_pregnancy_and_labour.aspx

http://www.hon.ch/Dossier/MotherChild/preg_changes/hormones.html

Urinalysis and other tests:

http://www.patient.co.uk/doctor/urine-dipstick-analysis

http://conovers.org/ftp/Urine_Dipstick_Testing__Everything_You_Need_to.33.pdf

http://www.medscape.com/viewarticle/763579

http://www.medscape.com/viewarticle/763579_4

http://www.mayoclinic.org/tests-procedures/urinalysis/basics/definition/prc-20020390